Paranormal Road Trip

Working on a Mystery

Larry Wilson

Copyright © 2020 Larry Wilson
First Edition: August 2020
All rights reserved.
ISBN-13: 978-1-7334631-2-6

ACKNOWLEDGEMENTS

To my family, friends, and colleagues for their patience during the completion of the writing of this book.

Contents

	Acknowledgements
6	Dedication
7	Prologue
11	The Petersburg Light
16	Mugged by a Ghost
28	Norb Andy's The Paranormal Pub
62	Legacy Theatre
78	The Monk of Elkhart Cemetery
91	Ghost Children
100	Doppelganger
105	Eye in the Sky- Cumberland Sugar Creek
124	Stranger in My House
152	Malvern Manor
179	Epilogue
182	About the Author
184	Books by Larry Wilson

DEDICATION

This book is dedicated to all who have experienced the strangeness that the supernatural and the unexplained have to offer. To my colleagues looking for answers to the mysterious and spectacular questions that the paranormal presents us.

Prologue

"All we can do, Scully, is pull the thread, and see what it unravels."
Fox Mulder- The X-Files

The above quote from the 1990s hit television series X-Files, is not only fitting for this book, it describes my curiosity and the way I approach the unexplained.

Or as many prefer to call it, the paranormal.

Things like ghost, cryptids, supernatural beings, and aliens. The type of things that are beyond the scope of normal scientific understanding.

Legends of ghost and monsters are discussed and often feared by society, whether it be in large cities, rural communities, or isolated civilizations.

Simple curiosity does not seem to justify investigating such things for twenty years like I have.

So, is there more to the story than meets the eye?

The answer to the question is, *"There is much more!"*

So, why do I continue to investigate the paranormal after so many years?

My answer: *Now that I have seen the magnificent, mind boggling and sometimes frightening things that the paranormal has to offer, I want to see more of them and discover what else is out there, where they come from and why they are here. Unfortunately, unraveling this mystery, is no easy task*

If I have learned anything since beginning my journey into the unexplained, it is, more things exist in Gods green acre than we have been told about.

For those of you who have not read my books, I will give you a brief introduction as to who I am, what I do, and why I do it.

My name is Larry Wilson. I am a former private investigator, turned paranormal investigator and author, who has been looking into the unexplained for twenty years now.

I am the guy people call when they think the boogeyman is roaming about their home, and many times they are right.

Most shy away from such things, while I seek them out, looking for answers to what these things are, where they come from and why they are here!

I have investigated the full gamut of the paranormal, from the mundane to the extreme, sometimes with a team of one or two investigators, but most times investigating alone.

In twenty years, I have experienced both the malevolent and the benevolent side of the paranormal.

For this reason, I treat what I do, as an endeavor and not a hobby or curiosity and pursue it with purpose and caution.

Most of us like to think that we know and understand our planet.

But when we see the mystery it has to offer firsthand; we find there are more questions than answers and the more we experience the strange things that the paranormal offers, the more perplexed the equation becomes.

Is not knowing the answers, a type of balance between nature and mystery, serving to keep us humble?

Perhaps our purpose is more about searching for the answers, rather than solving the puzzle

Why write Paranormal Road Trip?

Paranormal Road Trip is not about convincing you that the supernatural is real, because that is an impossible task for a book to accomplish.

The sole purpose is to share with you the things that I have witnessed and experienced over the last two decades as a paranormal investigator.

Things that have convinced me that ghost, cryptids, interdimensional beings, and other unknown creatures exist and intermingle with us.

I have experienced a reality that is bizarre and unexplainable. A reality that is both visible and invisible and one that interacts with us and knows who we are.

Yes Virginia, the Twilight Zone is real!

When I began investigating the phenomena two decades ago, I was not sure if ghost, monsters and the supernatural was real.

I wanted to believe, but the stories I read and the movies I watched, seemed to surreal to be true.

Eyewitness accounts by folks telling stories of coming face to face with ghosts and monsters, fascinated me as a young boy growing up in the small town of Petersburg, Illinois.

But the stories were coming from people I did not know and places I had never been. The odds someone like me would encounter such things, seemed impossible to a kid from a small town. Little did I know, all I had to do was go looking for them, and they would find me.

To date, I have witnessed seven supernatural beings, one of which I have seen on two occasions. Not in an obscure location, but in my own home.

In addition to witnessing supernatural creatures, I have recorded hundreds of unexplainable disembodied voices during investigations.

I can attest that the old saying, *"Truth is stranger than fiction,"* is accurate, when it comes to the paranormal.

Paranormal Road Trip chronicles some of the most fascinating and mind-boggling phenomena I have witnessed.

They are the type of experiences that keep me coming back for more after all these years.

Things that defy logic and cause me to re-think what I was taught in school. They are the type of stories, if I did not witness them for myself, would not believe them.

Hopefully, my stories will reassure those of you who have witnessed similar things that you are not alone in your experiences nor have you imagined what you witnessed.

To those who have not witnessed the strangeness that the paranormal offers and are not sure it exists, I hope this book will inspire you to begin your own examination into the unexplained.

I believe it is in our DNA to seek out the truth. But if I am wrong and not all of us share this trait. I assure you; I do, and plan to continue investigating the things that go bump in the night for years to come.

So, grab a hot cup of coffee or your favorite soft drink, and join me as I take you on a journey into the land of the strange and the home of the bizarre, the paranormal.

The Petersburg Light

My first encounter with the paranormal was on September 8, 1966. The date is easy to remember, because the night of my experience was the debut of the television show, Star Trek.

If you grew up in the 1960s as I did, with only three channels to watch, remembering the date of the debut of a television show is easy, but when you see something beyond the grasp of reality on the same night, you never forget it.

It was around 8:30 P.M. At the time I was living in Petersburg, Illinois. I remember that my mom was in the kitchen and my brother was doing homework. Dad was not home, as he was out visiting a friend.

Our living room was arranged with the sofa under a window. It faced the north and rested up against the wall.

I was bored, so I decided to look out the window, as I did on many nights. Climbing on the sofa, I kneeled on one of the cushions, with my elbows on the back of the sofa, and my chin resting on my interwoven fingers.

As a child of seven, I enjoyed looking out this window at night, to watch trains that stopped at a small depot on a hillside, about a mile north of our home at 604 North 4th Street.

It was dark, so I unclasp my interwoven fingers, cupping them to my face to block the glare from the living room light, that was reflecting on the window.

Pressing my nose against the glass, I remember the night as if it were yesterday. As soon as I looked out the window, I saw the stars, as it was a clear fall evening.

I noticed how large and strange the moon looked. It was the biggest and brightest moon I had ever seen. The color was an unusual, vibrant orange, its shape perfectly round.

Gazing at the bright orange moon for several minutes, its round shape and orange color mesmerized me. Then, something caught my attention.

Glancing to my left or northwest, and looking higher in the night sky, I saw the bright yellow, crescent shape of the moon.

I looked at the object which I mistook for the moon, then looked back at the actual moon.

The moon was much smaller than the object, which was the size of a half-dollar in the sky. I will never forget how perfectly round its shape was.

It was almost hypnotic as I stared at it. I could not take my eyes off it. After several minutes, it sped off at a high rate of speed and shot off toward the northeast, before coming to an abrupt halt.

It was now half the size that it had been moments before. Then once again, it shot off at a high rate of speed and came to a sudden stop. It repeated this behavior one more time, before it sped up and disappeared into the northern sky.

Over the course of the next few weeks, something strange happened to me. I developed a clairvoyance or foreknowledge, that even as a seven-year-old boy, realized was not normal.

Sudden thoughts and premonitions would come to me and the next thing that I knew, the visions would occur.

It was not something that happened all the time, but when it did, I would get an uneasy feeling that something was not right. A thought would come to me and the next thing I knew, before the day was over, the premonition would manifest into reality.

One example of this occurred when I had a vision that a train would hit someone. In the vision, I saw a four-door white car, being hit by a train, on the railroad crossing at the intersection where our house was.

Later that evening, I was lying on my parent's bed, staring out our south window. I could hear the train whistle blowing and could tell that it was nearby.

Next, I heard a car speed up and when I turned my head toward the east, saw the neighbor lady who lived around the corner from our house, speeding toward the railroad crossing in her white four-door car trying to beat the train. It was the same scenario that I envisioned a few hours before.

As I watched, the train struck the rear end of the car. The woman driver sustained injuries that resulted in her being taken to the hospital.

On another occasion, I was in school when the village fire alarm sounded. I looked over at one of my classmates and had the strange feeling that it was his house on fire.

I remember hearing him say, *"I sure hope it's not my house."* Later that afternoon, family members came to the school to pick him up, because it had, in fact, been his house that had caught fire.

After a couple of months, the visions stopped. But even as a child, I recognized what happened to me was peculiar. I believe the premonitions and strange light that I saw on September 8, were connected. But if they were, what did I see that night.

Was it extraterrestrial or supernatural? Was it closer to me than I realized and through some telepathy, opened my psychic awareness?

Was I caught up in a residual energy field by something from another world or dimension? If so, was it by accident or did some intelligent force know I was there and could see me, and as a result affected me.

In the summer of 2010, I attended a paranormal consortium in Springfield, Illinois. I was explaining what I saw as a child in 1966, to the Illinois Mutual UFO Network (MUFON) director, Sam Maranto.

As I explained my story, Sam would nod in an affirming manner, as if to say, *"Yes, I have heard a similar story before."* After I finished telling my story, Sam said. *"Larry, did you ever think when you saw the object speeding up at a high rate of speed, that maybe it wasn't traveling like you thought? What if I told you, that I believe you witnessed a wormhole in space, an inter-dimensional portal? Instead of it moving in the manner that you thought, what if when you saw the object getting smaller, you were seeing the wormhole closing up?"*

What Sam said made sense to me, and the more I thought about it, the more I agreed with his theory.

Even after fifty-four years, I still think about that night and remember it as if it were yesterday.

My mom passed away in 1994, but we discussed what happened that night even into my adulthood. Mom told me she knew I saw something that night because it left an impression on me. She told me that for the remainder of that September night and for days after, all I talked about, was the orange light.

Was what I saw, extraterrestrial, supernatural or as Sam Maranto said, interdimensional? I may never know the answer, but I do know that something happened that night that began my journey down the rabbit hole, called the paranormal.

I hope someone who reads this book, and who lived in the Petersburg area in 1966, may have witnessed the same event. If so, I wonder if they experienced premonitions like I did and caused them to pursue the paranormal.

In a later chapter, you will read of another encounter I had with a strange light that took place years later in a rural cemetery, that may or may not be related to the Petersburg incident.

Mugged by a Ghost

A very chilling incident occurred early in my paranormal investigating career. It took place at a cemetery in rural Christian County Illinois, in 2007.

The experience made a believer out of this former private investigator, who was trained to base conclusions on logic and tangible evidence, that ghost are real and roam graveyards.

Located near the small town of Palmer, Illinois is Anderson Cemetery. Or as some call it: *Graveyard X*.

The graveyard has been widely discussed by paranormal investigators and curious locals for years. It is a place that most prefer to steer clear of at night.

Strange sounds and voices of small children playing have been heard in the cemetery.

Unexplained floating lights have been seen during the daytime and at night. Anomalies appear in photos and digital thermometers have recorded icy cold temperatures in warm weather, with no scientific or logical explanation for the temperature fluctuations.

I personally experienced the extreme temperature variation which you will read about shortly.

To keep the location from others, books and internet postings labeled the cemetery, Graveyard X. This created a false perception that Anderson was a secret location.

In 2007, the stories and legends about the mysterious Graveyard, piqued my interest enough to compel me to find it, and conduct my own investigation into the legends.

Finding Graveyard X was my first challenge, because all I knew about its whereabouts was that it was in a secluded place, in rural, Central Illinois.

Fortunately, I uncovered a paranormal message board on the internet and found a posting requesting directions to the cemetery.

One of the message board members, posted a response, listing the latitude and longitude for Anderson Cemetery in Central Illinois.

I plugged the coordinates into *Google* and found a site that listed cemetery locations.

Unfortunately, the coordinates were for another Anderson Cemetery located in the same county.

Using the same message board, I eventually found directions to the correct Anderson Cemetery.

As it turns out, Anderson, or Graveyard X, is in rural Christian County only nine and a half miles from my home in Taylorville, also in Christian County.

So, it was right under my nose all the time. Oh well, so much for the intuition of a former private investigator!

My initial trip to Graveyard X was made during the daytime to make it easier to find and map out for a return trip to investigate at night.

Sunday March 4 was a clear day, with a temperature around thirty degrees and a slight breeze. To my surprise the graveyard was not the creepy, eerie place I expected. Instead, it was pleasant and well maintained.

However, on my next investigation, I would find out firsthand, that once the sun goes down, it is not so pleasant, and the legends were true.

Anderson is three miles from the nearest town, Palmer, Illinois. From Palmer, I headed north on fifth street, then traveled down a country road until coming to a green sign with an arrow directing travelers to the cemetery.

Turning left at the sign, I followed a narrow winding road for another mile and there it was: the *elusive* mysterious graveyard I had heard and read so much about.

I wandered about the cemetery conducting a pre-investigation, mapping out the grounds, taking photos, and using a digital voice recorder to see if I could record anything unusual.

Unfortunately, nothing eventful happened. The only activity I saw were several flocks of wild geese and a bald eagle flying overhead.

I took fifty to sixty photos hoping to capture a ghostly image, but only two of the photos had anomalies in them.

In the photos, a prism or rainbow color appeared. They were taken at high noon, so the sun was directly overhead. The rainbow effect may have been nothing more than glare or a reflection, but it puzzled me why the distortion only appeared in the two photos.

As far as I know, I do not have psychic abilities, but I had a strange feeling that something was there, and it was laughing at me. As if to say, *"Sorry, nothing is going to happen this afternoon, but come back at night and I will show you what I can do."*

I finished the pre-investigation and planned to return the following Thursday, to see what the graveyard had to offer at night.

I wanted to see if the tales and legends that made this place so mysterious were true.

Well, we should be careful what we wish for, because soon I would get more than I bargained for and experience what the forces at Anderson had to offer.

On Tuesday, March 6, I arranged with a close friend, who was also interested in the paranormal, to accompany me on Thursday nights investigation.

The plan was to arrive around 6:00 P.M., which would allow enough daylight to set up equipment and get a feel for the surroundings before darkness set in.

I knew that once the sun set; all the natural sounds of the wooded area surrounding the cemetery would change, becoming mysterious and creepy. Hopefully, getting familiar with the sounds of the location before sunset, would alleviate misinterpreting normal sounds for something paranormal.

As luck would have it, Thursday morning my friend informed me that he would not be able to go on the investigation, which meant for the first time, I would be investigating a secluded cemetery alone, at night.

I did not relish the thought of being alone in a secluded, rural cemetery with the reputation of Graveyard X.

At 5:30 P.M., I loaded my equipment and began the nearly ten-mile trek to the site.

For some reason, I kept thinking about the old Don Knotts' movie, *The Ghost and Mr. Chicken,* in which a mild mannered, wannabe newspaper reporter spends the night alone in a haunted house. The movie is one of my all-time favorites, and I felt much like the Don Knotts' character, Luther Heggs, who had cold feet at the thought of spending the night alone in a haunted mansion.

Although, this would be my first experience investigating a cemetery alone at night, it would not be my last, as I have done this many times since.

Over the last ten years, I primarily investigate hauntings alone, investigating the full gamut from secluded cemeteries, haunted houses and even several former funeral homes without fear. As a matter of fact, I prefer investigating alone.

When I arrived at Anderson, the sun was starting to set, and shadows were becoming more apparent.

I unpacked my equipment and began the short walk from the parking lot to the cemetery, I had no apprehension about the place, nor did it feel creepy.

Birds were chirping and there was a peaceful calm that gave a false sense of security about the place. This, however, would soon change.

One book I read, claimed the most active area in the cemetery, was in the oldest part of the graveyard at the north end.

The boundaries of the area described, were marked by a stone bench under a large oak tree, a tall monument, and a gravestone in the shape of an arch forming the graveyards own Bermuda Triangle of sorts.

I located the area during the pre-investigation and decided to concentrate my Thursday night investigation in this area.

The large oak tree was full of chirping black birds when I arrived. It reminded me of a scene out of the Alfred Hitchcock thriller, *"The Birds."*

Even though the chirping was annoying, it offered a sense of comfort, because if the birds were not on edge, why should I be.

Thirty minutes later, just before sun set, the birds suddenly stopped chirping. It was like someone flipped a switch and turned off the sound, because they knew something else was there besides me.

Not only did the birds stop chirping, the slight breeze that had been blowing, stopped. There was no movement of any kind in the surrounding woods either.

As darkness set in, the graveyard had a different feel about it, and the once peaceful feeling, turned to a feeling that I was being watched. Sort of a, *"calm before the storm,"* feeling.

Forty-five minutes passed and it was now pitch black in the cemetery.

I walked the perimeter of the triangulated area, looking and listening for anything out of the ordinary. Then stood in the same spot near my video camera for several minutes with my arms folded and my camera hanging around my neck.

From what I read, if I were going to experience ghostly activity, it would take place in this part of the cemetery. From the vantage point where I was standing, I had a good perspective of the entire area.

Everything was quiet when suddenly, I heard footsteps just a few feet in front of me and to my left.

When I heard them, I looked up expecting to see someone coming toward me, but no one was there.

It sounded like a person scurrying about. The grass was still crisp and frozen from the cold winter, so I could hear the crunching of grass as the footsteps scurried about.

Then, without warning, the footsteps walk past me. I felt the breeze caused by who or whatever it was.

It was the type of breeze you feel when someone passes you in a hallway or on a sidewalk. I not only heard them; I felt the presence of whatever it was as it passed me but saw no one.

I knew they were there, but they were invisible.

Pretending like I had not heard anything, I slowly turned around, and took a photo with my camera.

I reviewed the photo, but all that was in the picture were a few tombstones.

Even though the temperature was in the low thirties, my forehead was sweating like I was in a sauna and I was shaking like a leaf.

I stood quietly listening for footsteps or any movement.

After fifteen or twenty seconds of chilling silence, I heard them coming again.

Like before, they passed to my left. It was the same scurrying or shuffling sound. Once again, I felt the breeze as it passed by.

The hair stood up on the back of my neck, as I turned and attempted to take another photo. Unfortunately, the digital display of the camera flashed *low battery*, and the flash failed, so no picture.

The low battery message was puzzling, because I had just installed new batteries before arriving for the investigation. Although, it could have been caused by the cold temperature affecting the battery charge, after using the camera flash when I took the first photo.

Once again, I stood motionless listening for movement, thinking that who or whatever was there, was in front of me. I stood still, my hands trembling, listening, and facing toward the east, suddenly, someone or something, with great force, punched me in the middle of my back. The blow felt like a closed fist. But that was impossible because no one was there.

The punch was so hard, it caused me to lose my balance and stumbled forward. I was wearing a t-shirt, sweatshirt, and winter jacket that was unzipped halfway.

My jacket was puffed out in the back but when the punch landed, it hit hard enough that it pressed the jacket against my back.

An instant chill went down my spine and I could hear and feel my heart pounding in my chest.

It felt like someone was standing to my left, so I tried to take a photo, but again, the flash failed.

In shock, I stood still and listened for movement, wondering what would happen next, but nothing did.

I stood motionless a bit longer, then walked around the area for several minutes, when a sudden, empty feeling came over me.

I thought to myself, *"Hey, you are alone in a, secluded rural cemetery, and were punched by something you can't see and have no idea what its intensions* are. *Is it playing with you, trying to scare you or is it telling you, it's time to move along?"*

Since discretion can be the better part of valor, I decided to pack it up and call it a night.

Even though I had heard the rumors and read the stories about Graveyard X, I came to Anderson Cemetery with a healthy mix of skepticism and an open mind.

But never in my wildest dreams did I expect to be punched

by a ghostly entity.

What happened was the type of encounter I read about as a child, but never expected to experience for myself.

Whatever punched me, quickly made a believer out of me, that the stories and rumors about Anderson Cemetery were true, and ghost are real.

My experience was firsthand, so there was no denying the place was haunted.

In October of the same year, I returned to Graveyard X with paranormal investigator Ed Osborne. Ed was a very knowledgeable investigator and someone I had a great deal of respect for.

It was the Wednesday night before Halloween and was a cool, but not overly cold evening.

We had been in the cemetery for close to an hour, taking pictures, checking for electro-magnetic field (EMF) readings and monitoring temperature levels.

Nothing seemed to be out of the ordinary. The time was approaching 7:30 P.M. and the curfew for the cemetery was 8:00 P.M.

Since the county sheriff's department patrolled the cemetery, we decided to make a final pass through, and leave by curfew.

As we were making our final pass, I scanned the graveyard to check the temperature, using a digital laser-pointed thermometer.

Everywhere I scanned, the temperature reading fluctuated between forty-three and forty-four degrees with the average temperature at forty-four degrees Fahrenheit.

The readings remained consistent until I passed by the cement bench under the oak tree.

As I neared the bench, the temperature rapidly dropped. First it dropped below forty degrees Fahrenheit, then below thirty degrees, then twenty degrees.

The temperature continued to steadily drop until it finally reached a low of minus sixteen degrees below zero. This happened in less than a minute.

Neither Ed nor I could believe what we were seeing. To confirm there was not some type of equipment malfunction, or that the thermometer was set to Celsius instead of Fahrenheit. I shut the device off, turned it back on and reset it to Fahrenheit.

When I re-scanned the cemetery, just as before; the average reading was forty-four degrees everywhere I checked. Everywhere that is, except the spot next to the concrete bench, which was still below zero.

My hands were uncomfortably cold, and my nose felt like it was becoming frostbitten. Whenever I moved the thermometer more than four or five inches from the cold spot, the temperature would go back to forty-four degrees. But as soon as I moved it back, it would register below zero.

At 8:15 p.m., we decided to leave since it was after curfew and we were trying to be respectful of the cemetery rules.

The temperature was still minus eleven degrees Fahrenheit next to the concrete bench, when we left the graveyard.

My fingers and nose were so cold, when we got to my vehicle, I had to turn the heat on to warm them up, because they felt like they were frozen.

Based on an average temperature of forty-four degrees, at

one point, we experienced a sixty-degree temperature drop, which did not make sense.

I have been back to the cemetery many times since and have never recorded such an extreme temperature fluctuation again.

Ed and I continued to investigate together from time to time and without fail, the subject of our experience with the cold spot would come up. It was a mind-boggling experience and one that left a lasting impression on us.

Sadly, Ed passed away a few years ago, and is greatly missed.

Several years after experiencing the extreme temperature drop, I met longtime paranormal investigator Tim Harte, who I consider a friend and respect as a paranormal colleague.

During one chat, our conversation turned to Anderson Cemetery. Unbeknownst to me, Tim was part of an investigation and filming of a documentary at Anderson several years before my investigation.

As he was explaining some of the unusual phenomena that his team experienced, he mentioned an extreme temperature drop, that was recorded by three team members at the same time.

"It was crazy," Tim said. *"We recorded a temperature of thirteen degrees Fahrenheit in July, on an eighty-nine-degree day. The temperature was recorded using three different recording devices."*

I told Tim about the extreme temperature readings that Ed and I recorded in 2007. Tim had not told me the location where they recorded the readings in the cemetery.

When I told him, we recorded the below zero readings in

front of the cement bench under the large oak tree, his eyes grew wide. Because this is where they recorded the cold temperature as well.

Anderson Cemetery more than lived up to its reputation for me. Because as it turned out, the rumors of the place being haunted were not rumors at all but were haunting fact.

The footsteps and punch in the back were my first ah-ha moment, as a paranormal investigator, and proved to me, that ghost exist, and yes, they roam about in cemeteries as well.

Norb Andy's
The Paranormal Pub

Located at 501 East Capitol Street in the middle of the commercial section of downtown Springfield, sits a two-story, rectangular shaped, brick house.

Those who are acquainted with the historical side of Springfield, refer to it as the Virgil Hickox house, named for original owner Virgil Hickox.

Others more familiar with the night life of Springfield, it is better known as Norb Andy's Tabarin, named after the pubs founder and longtime owner, Norbert Anderson.

But history and nightlife are not all that 501 East Capitol Street is known for. To many, like me, it is simply known as haunted!

Virgil Hickox, a man with quite a resume, built the house in 1839. A Springfield Merchant and investor he lived in the house until his death in 1880.

Originally from New York State, Hickox moved to St. Louis, Missouri in 1928. After serving an apprenticeship, he worked as a journeyman carpenter until 1833 at which time he moved to the small Northern Illinois town of Galena, working in the lead mining industry.

It was in May of 1834 that Hickox moved to the Capitol City of Springfield, opening a general store located directly behind where the current building stands.

In addition to owning a general store, Hickox was involved

in banking, the railroad industry, and politics.

He served as the Chairman of the Democratic Party for many years having close ties with friend and presidential candidate Stephen A. Douglas.

Since the passing of Hickox in 1880, the building has built quite a resume, some of which is on the morbid side.

In 1895, the building became the location of the Sangamo Club, an exclusive men's club that was located at 501 East Capitol until 1911.

Branson's Funeral Parlor moved into the building in 1916 and remained at the location until 1926.

It was during the 1918 Spanish flu epidemic that the basement of the funeral parlor was used as the overflow to the county morgue.

As the story goes, people were dying so fast from the deadly virus, when someone died, their bodies were removed from the victims residence and transported to the funeral parlor, deposited down a shoot located on the West side of the building, where the bodies remained until they could be prepared for burial.

At one point during the epidemic, bodies in the basement overflow were stacked from floor to ceiling.

It is rumored that some of the bodies brought to the morgue were not dead but merely in a comatose state, later to be found alive.

From 1927 to 1938, the building housed several businesses including a flower shop and a combination sandwich shop and tavern called Cude's-Beer Stube.

The most well know business to occupy the basement level

of the building is Norb Andy's Tabarin, a nautical themed bar that includes a collection of knots used in sailing displayed above the fireplace.

Norbert "Norb" Anderson founded the business in 1939 and ran it for some forty years.

Subsequent owners have re-opened and closed the establishment since the days of Norb Anderson. Unfortunately, most have been short lived, lasting only a matter of months.

It was during one of the short-lived reopening's of the tavern, that I was allowed access for my most recent investigations.

The investigations took place on October 23, 2016 and Halloween night the same year.

The tavern reopened under the ownership of partners Dave Ridenour and Todd Gedaminski.

Between 2007 and 2012, I investigated the building on three occasions. All were somewhat uneventful, which would not be the case during the 2016 investigations, as the pub had a few tricks of the ghostly kind up its sleeve.

My decision to conduct a fourth investigation at Norb Andy's, stemmed from a simple trip past the old building after work one afternoon.

Capitol Avenue was not the shortest route home for me, but several times per month, I would take a detour on my way home, just to drive by the haunted landmark.

I am a bit obsessed when it comes to the paranormal, and locations like Norb Andy's attract me, like a magnet attracts steel.

I had driven by the building many times since my last investigation, each time thinking, what a shame such a beautiful landmark sat empty.

One afternoon was different when I drove by.

I noticed a sign near the entrance of the bar, indicating that it was open for business.

Virgil Hickox House - Norb Andy's

So, I decided to stop by and strike up a conversation with the bartender, hoping he or she would be receptive to talking about the haunted reputation of the building.

Plus, there is nothing I like better than talking about ghost while consuming a few pints of cold brew.

When entering the establishment, the first thing you notice is the nautical theme of the bar.

Particularly noticeable is a large ships wheel and framed collection of seaman's knots, hanging on the wall.

The dim lighting adds to the cozy appearance and atmosphere of the pub. Norb's has always reminded me of *Cheers,* the bar from the television series of the same name.

When I walked in, the tavern was deserted except for a young female bartender who greeted me.

I took a seat at the bar and we exchanged pleasantries.

She introduced herself as Sarah and asked what I would like to drink. For those who are curious, I ordered a tall Coors Light.

Wasting little time getting to my reason for stopping in, I handed Sarah my business card. I asked her if she knew the bar had a reputation of being haunted, which she did.

I explained to her that I had investigated the bar and Hickox house several times under the previous owners. Then proceeded to ask her if she had experienced anything unusual during her time working at the bar.

Her expression was enough to answer the question for me, but she verbally responded as well.

"Yes," she replied. *"As a matter of fact, I had something weird happen just a few days ago. Plus, I know that Todd, one of the owners, has had several strange things happen to him in recent weeks."*

"What did you experience?" I asked.

"Well" she said. *"I was working behind the bar like I am now when one of the employees brought a stack of freshly laundered towels and placed them on the bar for me to fold and put away. I was busy with a customer, so momentarily; I turned my back to finish serving them. When I turned around, I was shocked, because all the towels were neatly folded and stacked next to where I was working. No one was*

nearby and I had turned away for only a few moments. There was not enough time to fold and stack the towels in the short time my back was turned. It was freaky!"

"Stranger yet," she continued. "All of the towels were soaking wet and where they were placed was dry."

When Sarah finished her story, I confessed that the actual reason I stopped in, was to talk to the owners to see if they would allow me to do an investigation some night.

I explained my background as a private investigator and the length of time I had been investigating the paranormal.

Sarah indicated she was sure the owners would be receptive to an investigation, since both had experienced unusual activity and were open to discussing it.

"As a matter of fact," Sarah said. *"Dave, one of the partners will be here shortly, so you can ask him for yourself."*

While we chatted, a customer came in and a sat at the bar a couple of stools down from me.

Sarah was occupied with the customer, when a tall thin man walked in and sat down directly across the bar from me.

He was friendly, and we exchanged greetings from across the bar.

Call it fate or synchronicity, but no sooner than we exchanged pleasantries, the man began talking about an odd experience that he had in the Hickox House portion of the building earlier that day.

I could not believe what I was hearing. It was like he was reading my mind.

Whether fate or synchronicity, it was perfect timing.

Based on what he said, I assumed he was Dave the owner who Sarah had mentioned.

He began our conversation by asking if I knew the building was haunted.

"*Actually!*" I said. "*You won't believe this, but that is the reason I stopped in this afternoon.*"

"*Is that right*" he said with a surprised but interested look on his face.

With that, I grabbed my beer and seized the opportunity by taking a seat next to him on the opposite side of the bar.

After shaking hands, I introduced myself and handed him my business card.

I gave Dave a brief rundown on my background and explained I had investigated the building on several occasions and wanted to talk to someone about doing another investigation.

"*Well, you are speaking to the right man.*" he said. "*I'm Dave Ridenour owner of the building.*"

After introducing himself, Dave continued with the story he started to tell a few minutes before.

Earlier in the day, Dave and another gentleman were painting at the top of the stairs in the Hickox House portion of the building.

They had been working for several hours and nothing seemed out of the ordinary. Then discernible movement caught Dave's attention causing him to turn and look.

When he did, he saw a dark, shadowy mass moving along the wall near a doorway.

In shock at what he had just witnessed, Dave turned toward the other gentleman, who was looking at Dave.
"Did you see that?" The man asked Dave.

"I saw something," Dave replied. The two compared notes and agreed what they saw looked like a person moving about the upstairs.

"Darndest thing I ever saw." Dave said with a half grin, half puzzled look on his face. *"I knew the reputation of the place, but never expected to see anything like that."*

Upon accepting Dave's generous offer of a beer on the house, our conversation about the paranormal continued. He was curious about what I do as a paranormal investigator.

I explained my background and experience as both a private and paranormal investigator, then gave the Cliffs Notes version of how I conduct my investigations and explained a bit about evidence review.

Several minutes into our conversation, I figured that the time was right to asked Dave for permission to investigate.

He was receptive to the idea, with the stipulation that I also get permission from his business partner and co-owner Todd Gedaminski, to which I agreed.

Dave told me that Todd was gone for the evening but would be in the following day doing renovation work in the Hickox House.

"You should ask Todd about some of the weird things that he has experienced recently," Dave added. *"He's seen some really crazy stuff around here!"*

The following day, I returned to Norb Andy's on my lunch hour to talk to Todd as Dave had suggested.

It was shortly after 1:00 P.M. when I arrived, so the lunch crowd had thinned out a bit.

I asked a female employee working behind the bar if Todd was around. She pointed to the rear of the bar where a tall man wearing a baseball cap and yellow shirt was standing.

Wasting no time, I walked to the rear of the bar where the man was standing, introduced myself and explained the purpose for my visit.

I told him about the conversation I had the night before with Dave and asked if he had a few minutes to discuss the strange things he had experienced in the building.

"Sure," he said laughing." *But you'll probably think I'm off my rocker."*

I reassured Todd that I had seen enough crazy things over the years, that nothing he told me would make me think that he is crazy.

He began by explaining that before he became involved with Norb Andy's, he never gave any thought to ghost or the paranormal and had never really formed an opinion as to whether such things existed.

"But I'm a total believer now!" He said. *"You have to believe it because there's no other explanation for it. It's what we deal with here at Norb Andy's pretty much daily."*

Todd explained that he is a carpenter by trade and had been doing extensive remodeling work on the Hickox House portion of the building for several months.

"For ninety-six days I worked long hours in the building and never saw or experienced anything unusual. Then one morning that all changed." Todd said.

"I came in early to have coffee before I started work for the day. It was early, somewhere around eight o'clock. The only other person here was a guy who was doing some computer technology work for us. We were seated at that table," Todd said, pointing to a table directly across from the bar.

"We were talking and drinking our coffee when something behind the bar caught my attention. When I looked up, I saw this little blonde-haired girl running behind the bar toward the kitchen. She had long flowing curly blonde hair and was running and playing like she didn't know or care that we were here."

I asked Todd if she looked solid or more like an apparition. *"She just looked like a little girl running and playing and then disappeared into the kitchen,"* he explained.

Todd said the man, who was with him, must have seen his head and eyes following the little girl as she ran, because he looked at Todd and said.

"You just saw something didn't you?" To which Todd replied. *"You wouldn't believe it if I told you."*

His second experience happened under similar circumstances, once again while having morning coffee with the man at the same table.

Todd had his back to the front door and was looking toward the gaming room at the rear of the bar.

The room was used to drain the bodily fluids from corpses being prepared for burial when the building housed the overflow to the City Morgue.

Pointing toward the back room, Todd began explaining the story.

"We were just sitting and having coffee, when I saw something move at the back of the building. I looked up and saw this tall shadowy silhouette of a man standing in the doorway of the gaming room! He caught me totally by surprise. He was just standing in the doorway looking in our direction."

"It startled me!" Todd said.

"This ghost or whatever the heck it was, was huge! I turned toward the other guy, got his attention, and asked him if he wanted to see a ghost? But before I had a chance to show him, the man looked up and pointed toward the rear of the bar and said."

"You mean the guy standing over there?"

"He saw him too!" Todd said. *"I couldn't make out any facial features of the ghost, he was just this black silhouette of a man, and he was huge. You could not miss him. He was as tall as the doorway."*

We walked toward the doorway of the backroom as Todd continued describing what he saw.

At 6 feet 4 inches tall, Todd is well above average in height.

As he stood in the doorway of the room to show me how tall the figure was, I noticed that the top of Todd's head was a couple of inches below the top of the doorframe.

This meant that the shadowy figure they witnessed was at a minimum, 6 feet 6 inches tall.

Not long after Todd had witnessed the phantom girl and the shadow man, he had a third encounter, also in the bar area.

The incident took place during the daytime while customers were in the bar.

"I was standing in the dining area with a white bar towel slung over my shoulder," Todd began. "From behind, I felt someone put his or her fingers on my right shoulder and press down hard. Enough force was applied that it pushed my shoulder down several inches. Thinking that it was one of the staff playing a prank, I turned around to see who was messing with me. When I did, no one was there. It was one of the craziest things I have ever had happen to me," Todd said, shaking his head.

"Physically being touched and feeling the fingers of a hand actually pulling your shoulder down, makes you a believer pretty quick," He added.

Todd told of a more recent incident that occurred just a few days prior to my interview with him. He had been working the entire morning in the upstairs portion of the building, cutting wood trim, with a heavy commercial grade circular saw.

Deciding to break for lunch, he unplugged the saw, wound the cord up and placed the saw on top of an upside-down plastic milk crate.

Locking the door behind him, he left the upstairs portion of the building, then entered the bar from the Capitol Street entrance.

As he walked down the steps and into the bar area, one of his restaurant staff greeted him with a surprised look on her face. She questioned Todd by asking, "What the heck are you doing up there?"

"What do you mean?" Todd replied.

"Well, just before you walked through the door, there were several loud crashes that came from upstairs."

Based on the timing of when the girl heard the noise, Todd realized he would have already left the upstairs and locked the front door; so, it could not have been him that caused the noise.

Curious; Todd decided to head back upstairs to see if he could figure out what she heard.

The first place he checked was in the area that he had been working.

To his shock, he noticed the large heavy commercial grade saw he left a top the milk crate, was now positioned all the way on the opposite side of the room from where he left it.

"There is no way the heavy saw could fall off of the crate and end up all the way across the room." Todd exclaimed.

"It's simply too heavy. Even if I placed it too close to the edge of the crate and it fell off due to gravity, it could not roll across the room on its own. Plus, if it fell off the crate, there would have been damage to the floor, like a gouge or scratch, but there was no damage whatsoever."

If hearing Todd's accounts of seeing the ghost girl, the shadow man, phantom fingers touching him and a heavy commercial grade saw mysteriously move on its own wasn't enough to get my attention, he had one more experience to tell me about.

It would be an experience I would see repeat itself in a mere ten days, during my Halloween night investigation.

Todd began his story by having me follow him to a Juke Box, located on a wall adjacent to the bar.

He explained that in order to power the unit on, you must use a remote control, which is kept on a shelf located behind the bar.

The unit has two speakers, one located at the front of the bar and the other located to the rear of the dining area.

The volume level for each speaker can only be adjusted by using the remote control.

"The maximum we set the volume to, is level thirty, and we do so only when there is a large crowd. Otherwise we keep the level set at ten." Todd Explained.

"Several times our chef has told me when he arrived in the morning, the Juke Box was blasting away. It was not that I didn't believe him; I knew that for the unit to be powered on, someone has to have access to the remote." He continued

One morning after closing the bar the night before, it happened to Todd.

"I couldn't believe it!" He said.

"I know I powered it off the night before and walked out in silence. But when I came in that morning, the music was blasting so loud, I could not stand it. The volume for both speakers were set to level one hundred. It was the darndest thing I had ever seen." Todd explained, shaking his head.

When Todd finished telling me about his encounters with the paranormal at Norb Andy's, the conversation turned to discussing the particulars of scheduling an investigation.

Without hesitation, Todd agreed to the investigation.

Whether due to the construction, having customers and employees in the bar again or both, something stirred up activity taking place.

I knew from experience, the sooner I conducted the investigation, the better the chance I had to experience activity as well.

Because it seems that paranormal activity runs in cycles. Either a great deal of activity occurs or none at all.

Investigation
Friday October 21, 2016
Moon Phase - Waning Gibbous

I arrived at Norb Andy's at 9:30 P.M.

Todd was behind the bar and was talking to an attractive strawberry blonde, seated at the bar. I later found out she was his girlfriend, Terri.

Working alongside Todd was a female bartender, serving a drink to a man, who at the time was the only customer.

I remember thinking to myself, that a downtown tavern in the Capitol City, only having one patron on a Friday night, was not a good sign for business.

On the other hand, it was good for my investigation, since it would not take long to clear the bar after it closed.

I greeted Todd, then unveiled my plan of action for the night.

Since the bar was still open, I would not be able to investigate the basement until it closed for the night.

The Hickox house portion of the building was not in use, so I sat up my audio and video recording equipment in there, then when the bar closed, would set up equipment in the basement.

I knew from my earlier investigations that routine conversation from the bar does not interfere with audio recordings in the upstairs portion of the building.

At 10:00 P.M., I headed upstairs for the investigation.

Todd accompanied me and unlocked the outside entrance at the front of the building, since there is no direct access from the bar to the upstairs portion of the building.

After a quick walk through with me, Todd returned to the bar.

Audio recorders were in place throughout the building. Three in rooms on the second floor and three more in various rooms on the first floor.

One recorder was placed in a room on the first floor, with a newly constructed bar. I point out this room because several high-quality audio clips were recorded in the room.

I spent time in each room during the Hickox house portion of the investigation, which lasted from 10:00 P.M. to 1:30 A.M.

11:45 P.M.
I was on the first floor, in the room with the bar, seated on a barstool, listening for unusual sounds or movement.

One advantage of investigating alone in total darkness, is it enhances my sense of hearing, enabling me to hear subtle sounds I might miss, with the lights on or if accompanied by someone.

I had been sitting on the bar stool for a few minutes, when suddenly, I was overcome with the feeling that I was being watched.

Then without warning, I heard a loud bang coming from the next room. It sounded like someone slammed a window shut.

Earlier when we conducted the walk-through, I did not notice any open windows, so a window slamming shut, didn't make sense. Plus, it was October and chilly outside. If a window were open, I would have felt a draft from the cold air coming in.

I called out. *"Is there anyone here? If so, make that noise again."*

It was at this point; I started to get a weird vibe and felt a sudden change in the environment. I no longer felt I was alone.

Still seated on the bar stool, I listened for additional sounds. There was nothing at first, then a loud noise came from the second floor.

It sounded like someone pulled down a window shade and released it, causing it to recoil to the top of the window.

With flashlight and video camera in hand, I headed upstairs to see if I could figure out what had caused the noise.

Unfortunately, due to the clutter from years of storing unused items in the upstairs rooms, I could not determine what caused the noise or where it came from.

I walked around the upstairs listening for additional sounds, but heard nothing, so I headed downstairs and reclaimed my seat on the barstool.

Shortly after midnight I conducted the only EVP session of the night.

For those of you who are not familiar with the term, EVP. EVP is the acronym for *"Electronic Voice Phenomena,"* which are voices or sounds that are recorded by audio and video equipment, sometimes heard when recorded while other times they are not.

Paranormal enthusiasts have various theories as to what the voices are, including, disembodied voices of the dead, interdimensional beings, time travelers and demons to name a few.

During EVP sessions, I verbally lay out ground rules for the session.

For example, I will say something like, *"If there are any spirits or beings present, please answer my questions by speaking aloud. This way, I will be able to physically hear what you have to say, or my recording devices will record your answer and I will be able to hear your response later."*

Of course, unless a voice or sound is physically heard after asking a question, I don't know if a response was made until after returning home and reviewing the recorded audio.

On this night, the later would be the case, as I recorded several sounds and one clear voice that was a direct response to a question I asked.

A few days later while reviewing audio from the recorder near the bar on the first floor of the Hickox House, I found a response to one of my questions.

It was recorded shortly after Midnight, during the EVP session, after I asked the following questions.

"Were you married? What was your wife or husbands name?"

When asking the questions, the only sounds I heard, were the whistle of a nearby train, and a sliding sound, caused by my foot that was on the floor, while seated on the barstool.

In the recording, you hear my foot making the sliding sound, followed by a male voice that whispers, *"Nicole."*

Was the voice a spirit saying a random name or was it answering my question and telling me that he was married to someone named *Nicole?*

Whatever the case, it shows an intelligent being was present, who heard my question. In my opinion, the timing of the answer is too precise to be coincidence.

In addition to the voice whispering, *"Nicole",* I recorded the banging sound, I heard earlier coming from the upstairs.

Another compelling piece of evidence was recorded shortly after the *"Nicole"* voice.

I recorded what sounds like someone snapping their fingers, immediately followed by human sounding whistling.

It was as though someone were trying to get my attention or was curious to see if I could hear the sounds they were making, which I could not.

It was approaching 1:30 A.M., and all was quiet in the Hickox House. The bar closed at 1:00 A.M., so I headed downstairs to set up additional recorders.

When I entered the bar, Todd, the female bartender, and Teri, were having a conversation as all the customers were gone for the night.

Todd asked how things had gone upstairs, so I gave a brief rundown, explaining the noises I heard.

I then set up additional audio recorders. One across from the bar, one near the Jukebox, and one in the backroom where Todd witnessed the shadowy figure in the doorway.

I placed a video camera equipped with infrared capabilities at the front of the bar.

Once the equipment was set up, I joined Todd, Terri and the bartender further explaining the noises I heard upstairs.

During the conversation, the female bartender told me how she always gets an uncomfortable feeling in the upstairs portion of the building.

She explained she had never seen or experienced anything to cause the feeling, but it was overwhelming.

At 2:00 A.M. the bartender decided to leave for the night. Since it was late, Todd escorted her to her car.

It was while Todd was walking the bartender to her car that Terri and I witnessed something we could not explain.

We were seated at the corner of the bar with our backs to the front door, having a conversation, when the wall near the door leading to the kitchen, lit up in a bright orange glow.

The glowing light, lasted three or four seconds, then vanished. All the lights in the bar were off, so the glowing light was clearly visible to us.

We turned and looked at each other, both saying. *"Did you see that?"*

I got up from my seat and hurried to the other side of the bar. I did not see anything that could have caused the glowing light.

After comparing notes, we concluded we both saw the same thing, a glowing orange light.

When Todd returned, we told him what we saw.

Unfortunately, the remainder of the night was uneventful, so, at 4:45 A.M., I packed up the equipment and we ended the investigation.

As I drove down Route 29 heading home, I reflected on the events that took place. *"Nothing earth shattering had occurred"* I thought to myself.

I heard a couple of unusual sounds, but after all, it is an old cluttered building.

The glowing orange light was harder to explain away, as I did not have a logical explanation for what caused it.

I was not disappointed at the lack of activity during the investigation, because over the years, I have found that it is usually the uneventful nights, when the best audio evidence is recorded.

As it turned out, I recorded the *"Nicole"* EVP which was a direct response to my question and recorded the snapping and whistling sounds. Both of which, were good evidence.

Plus, little did I know as I was driving home, that in only nine days, I would get another chance to investigate the building, and this time, it would show me what it could do!

Wednesday October 26, I received a Facebook message from local television personality Lindsey Hess.

In her message, she asked me to call her and provided her telephone number so that I could do so. She indicated that she wanted to do a story related to the paranormal to be aired on FOX 55 for Halloween.

So, after work, I called Lindsey from my car to get further details about what the focus of her story would be.

She told me that she had been discussing the possibility of doing a segment on the paranormal with her friend and colleague Chris Neal, who works behind the scenes at the station in production.

It was Chris who suggested she contact me. Lindsey went on to explain that I knew Chris, but I knew him by another name.

As it turned out, I did know Chris by another name, and through his other profession as a local radio personality.

Chris is better known as Jammer, who worked for Springfield radio station, 99.7 KISS FM, which is now 99.7 THE MIX.

I had been a frequent guest discussing my paranormal adventures; on KISS FM's Morning Grind radio show hosted by another radio personality and friend Bondsy.

It was during one of my appearances that I met Chris. So now the connection to Chris made sense.

Lindsey explained that what she wanted to do, was tag along with me on an investigation at a place believed to be haunted and record the investigation.

She would present the results of what we found and her experience on their Halloween broadcast.

The only night Lindsey had available to do this was Sunday night October 30th, just four nights away.

I explained I would be happy to take her somewhere, but our main obstacle would be time.

By this I meant, that after investigating, it takes time to review all the raw audio and video that is recorded, to see if any evidence has been obtained.

Conducting the investigation, the night before the broadcast, would not allow enough time to do this.

Since we would not have enough time to review the evidence, I presented another option to Lindsey.

I told her about my recent investigation at Norb Andy's and the audio evidence I recorded, including the *"Nicole"* EVP and the snapping and whistling sounds.

The option I presented to Lindsey was to conduct the investigation at Norb Andy's, so she and her cameraman could tag along with me, to see and experience what a paranormal investigation is like.

Since there would not be enough time to review the audio and video from our investigation before the broadcast, I would allow her to use the *"Nicole and snapping and whistling"* EVP's I recorded during the first investigation, for the newscast. Then during the newscast, she could explain the audio was from a recent investigation and that she would follow-up once the evidence was reviewed from our investigation.

I contacted Dave and Todd to get permission to do the investigation, and to see if they would be willing to go on camera with Lindsey, to tell stories of the ghostly activity they witnessed.

Permission was granted and both agreed to interviews.

Investigation
Sunday October 30, 2016 -
(Moon Phase New Moon)

On Sunday evening, October 30, I arrived at Norb Andy's at 6:30 P.M., the pre-arranged time Todd and Dave had agreed to meet me.

The bar closed at 6:00 P.M., so there would be no noise distractions from customers to contend with during the investigation.

Lindsey would arrive at 7:00 P.M. to conduct her interviews with Todd and Dave. When finished, she would head back to the studio to anchor the 9:00 P.M. news broadcast, which would give her plenty of time to make it back to Norb Andy's to begin our investigation at 10:00 P.M.

While waiting for Lindsey to arrive, I played the audio clip that I had recorded from the previous Friday night's investigation for the two owners.

Both were amazed at the clarity of the whispery voice saying *"Nicole!"* Neither knew of a connection between the name or anyone associated with the bar or building named Nicole.

Historical research I conducted did not turn up a connection either.

Being the home of the former county morgue, there are many names connected to the building, that historical records would not reflect.

At 7:00 P.M., Lindsey arrived. With her, was a cameraman from the station and her colleague and friend, Chris Neale.

After exchanging pleasantries, Lindsey conducted the interviews with Dave and Todd as planned.

When she finished, we went over the game plan for the night, which included placing audio and video recorders throughout the building, conducting several EVP sessions and staking out the place, in anticipation of catching a glimpse of the strange activity that takes place.

Staking out a building, is the same basic method of investigating that I used when conducting surveillances as a private investigator.

So basically, sit tight and wait to see what unfolds.

It is amazing what one can see and hear when sitting quietly in the shadows of a darkened building or secluded location.

After I finished explaining the plan for the night, it was time for Lindsey to head back to the station to prepare for her 9:00 P.M. newscast.

While the television crew was away, I placed audio and video recording equipment throughout the building.

I chose locations where unexplained sounds and voices were experienced during the previous investigation, as well as areas where Todd and Dave experienced activity.

Lindsay and Jammer returned shortly after 10:00 P.M. All the equipment was in place and I could see by the enthusiasm and excitement in their voices, they were looking forward to spending the night in a haunted building.

As is the case for most who are about to embark on a paranormal experience for the first time, I could see a bit of apprehension to go along with their excitement.

It is the not knowing if some ghostly being is watching your every move and standing next to you. I see this when taking someone with me who has never been on a paranormal investigation before.

Their eyes are in constant motion, their movements slow and calculated, almost as if expecting that an unseen hand will reach out and grab them from the shadows.

The investigative team for the night would consists of Todd, Lindsey, Jammer, and me. Dave would not be part of the team, as he left after his interview with Lindsey.

We began our adventure in the Hickox house. The bar and the house have separate entrances, so each time we moved our investigation from the bar to the house, Todd locked the door behind us, so no one else had access to that part of the building.

You will see the importance of this information, later in the chapter.

We turned all the lights off and used flashlights to find our way around.

I thought it best if the four of us remained together during the investigation, so when I reviewed the audio, it would be easier to determine if something said was one of the team, or the disembodied voice of a ghost.

Early on the investigation was uneventful, or so it seemed. Because it was not until a few days later when reviewing recorded audio, that I discovered we had activity going on right under our noses, as several clear EVPs were recorded.

But I am jumping the gun a bit, as the EVP's will be discussed later in the chapter.

However, during the investigation, two unexplained events occurred, including a cool but creepy event that took place at exactly, 4:03 A.M. It is an experience the investigative team will tell about for years to come.

The first unexplained activity, occurred a few minutes before 2:00 A.M.

We were in the hallway on the east side of the Hickox house. Todd was several feet away, facing Lindsey, Jammer, and me.

Our flashlights were off, so it was completely dark.

Todd was re-telling the story about being grabbed by his shoulder while working in the bar.

Suddenly, I heard what sounded like a woman talking. The voice came from the south or directly behind the group.

Immediately I asked. *"Did anyone just hear a voice coming from behind us?"* To which, both Lindsey and Jammer replied. *"I did!"*

Todd, who was talking when the voice spoke, said, *"I didn't hear anything."*

We agreed that the voice was female and was only a few feet behind us. We couldn't hear what was said, since the voice spoke while Todd was telling his story.

Unfortunately, as many times is the case during paranormal investigations, even though the voice was loud enough for us to hear, it was not recorded by our audio recording devices.

We did, however, record two female voices in the Hickox house during our investigation, which I will discuss in the *"Summary of EVP's"* section of this chapter.

We spent another hour in the Hickox house, but never heard the voice again. Since all was quiet, we decided to lock up and move next door to the basement tavern to continue the investigation.

Little did we realize, in less than an hour, we would witness an event, we will never forget.

Since the tavern portion of the building is relatively small, we did a short walkthrough of the backroom and the east hall storage area, then camped out at the bar.

Lindsey, Jammer, and I were seated at the northwest corner of the bar, with Lindsey to my left, and Jammer to my right. Todd was behind the bar and leaning on it with his forearms.

After discussing some of the peculiar things that Todd had witnessed since becoming associated with the tavern, we decided to do what I call, *"quiet time."*

Quiet time is like conducting a stakeout for unexplained noises and voices. Then if something is seen or heard by one of the team members, we investigate, and try to find the source of the activity.

Before we started quiet time, I commented to Jammer, how it felt like someone was watching us, from the doorway at the south end of the room, to which he agreed.

All the lights were off, so the bar was very dark.

As I turned my head toward the left, in the direction of Lindsey, something caught my eye in the obscure lighting.

I saw a powder blue light, reflect on the wall just behind her. It was like the orange glow I saw the week before that came out of nowhere.

Before I could say anything, a pinkish light, reflected off Todd, who was still standing behind the bar in front of me.

Next, a bright yellow light lit up the bar area, illuminating the entire investigative team.

At first, I could not tell where the source of the light was coming from, then in an excited voice, Jammer said, *"Look, the jukebox is turning on!"*

Simultaneously the group turned to the right and looked. Sure enough, the jukebox had powered on all by itself. The pastel colors I had seen reflecting on the wall, were coming from the now brightly lit jukebox.

Then, without warning, a song began to play.

When the days are cold
And the cards all fold
And the saints we see
Are all made of gold
When your dreams all fail
And the ones we hail
Are the worst of all
And the blood runs stale
I want to hide the truth
I want to shelter you
But with the beast inside
There's nowhere we can hide
No matter what we breed
We still are made of greed
This is my kingdom come
This is my kingdom come
When you feel my heat
Look into my eyes
Its where my demons hide
Its where my demons hide
Don't get to close
Its dark inside
Its where my demons hide
Its where my demons hide

The song?

"Demons", by the group, Imagine Dragons.

The jukebox is an iTunes jukebox that can play thousands of music selections. So, what are the odds of this song, playing in a haunted bar being investigated for ghosts?

My guess is, astronomical!

Do I think that a demon or some type of malevolent spirit played this song to scare us? No, not at all.

But I do believe that one of the spirits or beings that haunts Norb Andy's, has a sense of humor.

A few days after witnessing the jukebox turning on by itself and playing its own selection of music, I researched iTunes jukeboxes and how they work.

Although, playlist can be programmed using any cell phone having the model and serial numbers of the jukebox, one thing they cannot be programmed to do, is turn the machines on or off.

To do this, the remote for the unit, must be used.

The night I witnessed the jukebox playing on its own, the remote was stored on top the refrigerator behind the bar and no one was anywhere near it.

What we saw, validated the experiences of Todd and his cook, who found the jukebox playing several times when they arrived for work in the morning.

Unfortunately, the battery for the camera I had pointed in the direction of the jukebox ran down before the box began to play. But there were four eyewitnesses who saw it.

The ghost of Norb Andy's must have saved its best trick for last, as nothing else was experienced by the team the remainder of the night.

I was scheduled to appear on a morning radio show for Halloween at 6:00 A.M., so we began breaking down the equipment at 5:00 A.M., and called it a night.

Lindsey and Jammer were lucky. Because it is rare for someone tagging along on their first investigation to be fortunate enough to witness something strange like a jukebox playing on its own.

Not to mention they heard a disembodied voice as well which in and of itself will make a believer out of you that ghostly things exist.

If you ever have the chance to meet Lindsey or Jammer, asked them about the night they spent in the haunted Norb Andy's and watch their eyes light up.

What they witnessed are the type of things that keeps you coming back for more of the strangeness that the paranormal has to offer.

EVP Summary

During the investigation, EVP's were recorded as early as 8:38 P.M. and as late as 2:55 A.M., in the Hickox house. Nothing was recorded in the bar area.

When the EVP's were recorded, no one was in that part of the building.

A couple of the recordings have me shaking my head a bit. I would place them in the category of the bizarre, when compared to most EVPs I have recorded over the years.

As I mentioned, when the 8:38P.M., EVP was recorded, no one was in that part of the building, and the only entrance, was locked.

The best way to describe what we recorded; is it sounds like a witch laughing. The type of laugh you would expect a cartoon character to make during a children's Saturday morning cartoon show.

The laugh is a loud and clear, *"Ha-ha-haaa!"* The last part of the laugh has emphasis on it and is drawn out. It is totally bizarre.

A second and equally bizarre voice was recorded shortly after 11:00 P.M.

You will recall I mentioned earlier, that during the October 21st investigation, I heard what sounded like a window slamming shut, but when I searched for an open window, I could not find one.

During the October 30th investigation, the team and I were discussing the sound of the window slamming that I heard during my first investigation. As a result of our conversation, we decided to try and recreate it.

To do so, Todd opened a window that was only a few feet from where I had been standing the night, I heard the noise.

He tried to recreate the sound, by forcibly shutting the window. When he did, it sounded nothing like what I heard and recorded on the 21st.

I discovered upon reviewing audio evidence, that during Todd's re-creation, another bizarre EVP was recorded.

As soon at the window slams, the voice of a female child can be heard singing, *"You're going to hell."*

The rhythm and lyrics of the short tune reminds me of something that you would hear in a Stephen King movie.

Although conjecture, based on the timing of the voice and Todd shutting the window. I wonder if she was shaming Todd for closing the window or shaming us for looking for ghost?

Several other EVPs were recorded throughout the night, but the next disembodied voice that can be understood, was recorded at 2:50 A.M.

It was recorded after we left the Hickox House and Todd locked the door which makes it even stranger.

I place the EVP in the category of bizarre, due to the sequence of events that can be heard both before and after the voice speaks.

The first thing heard in the recording is the sound of two footsteps, that abruptly stop. As soon as the footsteps stop, a male voice calls out *"Danielle."* Then, there is a squeak that sounds like the hinge on a door followed by the sound of the door shutting, then silence.

It was not until I played the recording for Lindsey and Jammer, that I found out from Lindsey, that she had a close family member who had been tragically killed.

The name of her loved one, was Danielle.

There is no way to know if the Danielle that the voice was calling out to, has a connection to Lindsey's deceased family member or if it is simply a coincidence.

With the number of dead bodies that were brought to the building when it was the overflow to the County morgue, it is possible that a lost soul was looking for another Danielle who passed and was brought to the building.

One of the clearest EVPs recorded during the night, was recorded only five minutes after the Danielle EVP was recorded.

At 2:55 A.M., a very clear female voice says, *"Come close."*

The voice is a beckoning call for someone to move closer, but since none of us were in the Hickox house at the time, I would assume that none of us were being summoned.

Mysterious bangs and phantom voices heard coming out of the darkness. A juke box powering on by itself, playing a song titled, *"Demons."* Footsteps, doors closing, and names being called out, all recorded.

These are the type of things that you would expect to find in a haunted location, and as you have just read, we did.

If experiences like this are not enough to convince one that a location is haunted, then I am not sure what will.

I know firsthand, what Lindsey and Jammer experienced, made believers out of them, that the paranormal is real.

Because just as Todd said to me the first time that I met him. "

"You have to believe it, because there's no other explanation for it."

I agree with you Todd, and it is places like Norb Andy's that keep me coming back for more and make paranormal road trips worthwhile.

Legacy Theatre

There are places I investigate, that I can tell are haunted as soon as I walk in. I am not psychic, but when I enter such a place, that sixth sense that we all possess, seems to kick in.

These places are not necessarily spooky, but the atmosphere has a feel to it that someone or something is there. The environment is different.

It is a feeling of a presence from the past that still lingers. If you venture off by yourself, you can feel eyes watching you.

Over the last twenty years, I have found this type of place to be varietal gold mines for recording disembodied voices.

Springfield, Illinois Legacy Theatre at 101 East Lawrence Street, is the type of place I am talking about.

Located in the heart of Illinois, Springfield serves as the hub of the state's political activity.

Springfield became the state capital in 1839 with the help of a young lawyer named Abraham Lincoln, who lived in Springfield until 1861, when he left to become the 16th President of the United States.

Many believe that citizens from days gone by, still roam the streets and buildings of Springfield; appearing out of nowhere to haunt those whose paths they cross.

The quaint and friendly atmosphere of the theater gives the

feeling that you have taken a step back in time and perhaps when you enter the building, you do.

The Springfield Theatre Center, as it was originally called, was the venue used to showcase live performances for the Springfield Theatre Guild.

On November 8th, 1951, the Theatre Center opened its doors with a performance of the Broadway show *Born Yesterday*.

The business received congratulatory telegrams from famous celebrities, such as Bob Hope, Bing Crosby, and Broderick Crawford.

One of the members of the Theatre Guild was actor Joe Neville.

According to some, Joe was odd and at times arrogant. Some cast and crew members did not like him.

In addition to his arrogance, Joe had a mysterious side to him as well.

It was rumored that he previously acted in England under a different name, but being a talented and dedicated actor, his past was overlooked by his fellow actors.

In 1955, Joe was given the lead role in the play, *Mr. Barry's Etchings* and all was well, or so one would think.

Unfortunately, things were not as they seemed. After a dress rehearsal for the opening of the play, Joe returned home and committed suicide, overdosing on pills, never performing in the show.

It was determined the reason Joe ended his life was due to an audit at his place of employment. Funds had been misappropriated and Joe was the likely suspect.

But as they say, *"the show must go on,"* and his role was assigned to another actor one day before the show was set to open.

Today, many who frequent the Legacy believe that Joe's spirit continues to linger inside the theater.

Reports of paranormal activity began almost immediately following Joe's death and continue to this day.

Actors and stage crew have reported strange sounds, such as doors opening and closing on their own, lights turning off and on without reason. Costumes and tools disappearing only to be found later, folded, or placed in an area which had been previously searched.

Some claim they have seen Joe's spirit wandering the Theatre Center. Most of the sightings occur prior to the opening of a new show.

Today, the Theatre Center has a new owner and a new name. The Legacy Theater.

Owner Scott Richardson has started a new era as he greets the faces of fresh guests who come to be entertained. But not all the entertainment takes place on the stage.

Strange encounters have led some to believe that spirits from the past continue to linger, refusing to move on to their final resting place, preferring to remain and frequent the quaint and friendly atmosphere of The Legacy.

Owner Scott Richardson has experienced some of the strange phenomena that occurs on a regular basis, firsthand.

I met Scott in early June 2011, just weeks after he purchased the old theater.

I had been trying to get into the Theatre Center for months but was unable to find the right person to authorize an investigation.

Fortunately, my good friend Lynn Puls, a local hairstylist and owner of The Hair Shanty, was asked to help style wigs for a play taking place that July at The Legacy.

Although Lynn had never met Scott before, he asked her to meet with him to discuss hairstyles for the play. During the meeting with Scott, Lynn mentioned that her friend, Larry Wilson, was interested in conducting a paranormal investigation of the theater.

Lynn asked Scott if he knew who to contact to allow access to the building for an investigation and Scott replied, *"As of a few days ago, me!"*

I met with Scott during the first week of June and he gave me a ninety-minute tour of the theater. During the tour, Scott described some of the strange occurrences that he had encountered since he came into ownership of the building.

When Scott was considering purchasing the theater, he went on several walkthroughs of the building. During each of the visits, he noticed that the building gave off a depressed or negative feeling. But, after he purchased it, and walked in the door for the first time, as the new owner, he noticed that it gave off a completely different vibe.

It had a feeling of relief, as if the building knew that it was about to get a face lift and soon would come alive again with performances and spectators.

The theater sat idle for several years and needed a lot of work to get it in condition to once again host performances.

There was so much to do that Scott was not sure what needed to be done first.

He decided to start with the landscaping, to give the outside a fresh look.

Perplexed at where to begin, he contacted a local landscaper and arranged to meet with them the following day.

"I locked the building before going home for the evening and took the only key with me." Scott explained.

"The next morning, when I unlocked and entered the building, I found the original building plans, rolled up on my worktable. They were not just any plans; they were the original landscaping plans."

Scott said that he had been through all the paperwork that was left in the building, and the original plans for the exterior, were not included with the paperwork.

"No one, but me, had access to the building. Where the plans came from, is still a mystery." Said Scott.

This was just the beginning of strange happenings.

One day, Scott was doing some repair work on the stage using a hot glue gun.

"I laid the glue gun down next to me and turned away for just a split second. Scott Said.

"When I turned back, the gun was gone and was nowhere to be found. Later, I was getting ready to leave for the night when *I noticed something behind the last row of seats in the auditorium. Underneath the seat, with the cord neatly wrapped around it, was the glue gun. To this day, I have no idea how it got there."* Scott said shaking his head.

Another evening, Scott was alone in the building when he heard what sounded like someone taking a handful of nails and throwing them in the air on the stage.

"I heard the pinging sound of nails landing on the stage, but when I checked it out, nothing was there." Explained Scott.

Stage and Auditorium

Investigation
June 17, 2011 -
(Moon Phase Waning Gibbous)

On June 17th, 2011, I investigated the Legacy for the first time. Accompanying me was a paranormal enthusiast named Jay, who was on his first investigation.

During the investigation, a few showers and thunderstorms moved through the area which added a spooky atmosphere to the night.

We arrived at the Legacy at 7:00 p.m. with Scott arriving a short time later.

Scott decided to stay for a while and do some spackling work to prepare walls for painting. As he went about his business, Jay and I did a walkthrough of the building, recording baseline temperature and electromagnetic field (EMF) readings.

Many paranormal investigators believe fluctuations in EMF, is a by-product of the presence of spirits.

Unfortunately, homes and buildings like the Legacy have a lot of electrical wires running in walls and underneath floors, which cause fluctuations in EMF, making the data unreliable in the detection of ghost and supernatural beings.

For this reason, I do not spend a lot of time monitoring electromagnetic field readings.

However, monitoring temperature is a different story. I have more faith in the use of temperature fluctuations in ghost detection, as I have personally experienced extreme temperature variations during investigations. You read about one such incident that took place at Anderson Cemetery earlier in the book.

For the most part, during the Legacy Theatre investigation, the temperature was seventy-five degrees upstairs and slightly cooler in the basement.

Scott left at 9:00 P.M., so Jay and I set up video cameras and strategically placed digital audio recorders throughout the building.

For the most part, the night was quiet, and we did not have any personal paranormal experiences.

However, around 11:00 P.M., I was standing in the doorway of an old dressing room in the basement that Joe Neville may have used at one time.

Jay was in the dressing room conducting an EVP session. Suddenly, I felt someone or something tickling my back. My first thought was that something ghostly was touching me, but when I turned around, I discovered a bat was fluttering on my back!

As the night progressed, we were disappointed in the lack of activity. After all, a place with the reputation of the Legacy tends to get your hopes up.

At 4:00 A.M., we completed our investigation and began to break down our gear. Even though the night was uneventful, we had hours and hours of recorded audio and video to go through, so we hoped our time and effort would be fruitful.

As it turned out, it was. Because upon reviewing our evidence, I found audio that proved we were not alone during the investigation, and had more company, than a pesky bat!

EVP Evidence

One of the strangest EVPs was of someone speaking with a foreign accent. When it was recorded, Jay and I were having a conversation about the investigation.

Although Jay's heritage is from India, he was born and raised in Central Illinois, and does not have an accent. During our conversation, after Jay speaks; you hear a loud and clear voice with an Indian accent say, *Creloza*.

We do not know what it means or if it is possibly a person's name. I have talked to people of Indian descent and they have never heard the word before. We also used online language converters without success. So, the clip remains a mystery.

The next EVP is one of my favorite EVP's that I have recorded.

We were in the basement near what was once the concession area. I had placed a digital recorder near a door on a folding metal chair.

Unexpectedly, a sump pump turned on in a nearby room, causing a lot of noise that I knew would create a problem for our audio recorders.

So, I decided to close the door to muffle the sound. To do so, I needed to move an extension cord that was draped over the top of the door, to close it.

I had a video camera in my hand, so I laid it on the chair next to the audio recorder and shut the door. As I shut the door, the sump pump turned off.

When I reviewed audio from the recorder on the chair, I found that I had recorded a voice that seemed to be talking to me.

In the audio clip, as soon as I close the door and the sump pump turns off, a clear male voice whispers, *"You left shit there,"* as if to remind me that I had left my equipment on the chair.

The recording is one of the clearest EVPs I have recorded

in the last twenty years.

The spirit seemed to be interacting with me, which means that it knew I was there, thus it was an intelligent spirit.

In case you are not familiar with what I mean by an intelligent spirit or haunting. An *intelligent haunting* is when a spirit or entity interacts with the living, having an intellectual awareness about it.

The next EVP was also an intelligent spirit. The voice sounds exactly like the voice who declared, *"You left shit there."*

When the EVP was recorded, Jay and I were having a conversation about our disappointment with the lack of activity during the investigation.

In the clip, you hear me say, *"It's really calmed down out there."* Referring to the thunderstorm that had moved out of the area. To which Jay jokingly replies. *"Well, at least we experienced the bat flying around."* Referring to the bat that had bumped my shoulder.

Immediately after I say, *"It's really calmed down out there."* A clear whispery male voice says, *"That's your own conclusion."* So, it seemed that a spirit had a difference of opinion!

Two compelling EVPs of singing were recorded forty-five minutes apart. The singing was not just ordinary singing but sounded like a man and a woman rehearsing for a play or performance. It is simply amazing.

Even though the June 2011 investigation did not produce personal experiences, modern technology allowed us to capture evidence, of an invisible world that cannot be seen with the naked eye, but exists, nonetheless. The clear audio evidence offers proof that activity is taking place at the Legacy

Theater and gives credence to the stories that have been told for years.

I conducted a second investigation that same year, in which physical manipulation of a light was witnessed.

Investigation
October 31, 2011
Waxing Crescent

The June investigation may have lacked personal interaction with the ghost of the Legacy, but Halloween night of the same year would offer up the type of unexplained interaction that keeps a twenty year investigator coming back to see more of what the supernatural can offer.

The incident that I refer to, involved a motion sensor security light in the basement.

Once again, Jay accompanied me on the investigation along with three paranormal colleagues and investigators named Jamie, Chris, and Tim.

Tim was along on the investigation to use specialized equipment called MESA which stands for *Multi-frequency Energy Sensor Array,* that measures a variety of energies associated with hauntings.

The equipment has been used at well over one-hundred-fifty locations. The system collects data on infrared, visible and ultraviolet light intensities, natural and artificially generated electromagnetic fields, gamma ray, radiation, galvanic skin response of a human subject, infrasound, and vibration.

They also deploy still and video cameras and audio recording to document haunt phenomena. If you would like to find out more about MESA, you can do an intranet search using the key words, *"MESA Project.*

Unfortunately for Tim, he was monitoring his equipment on the stage area when the unexplained interaction that I am about to tell you, took place.

The remarkable incident stemmed from a motion sensor light in the basement that was interfering with an infrared camera we were using in the old dressing room area.

The light from the security system was a problem, because, it caused an overexposure for the infrared camera which is designed to film in the dark.

Even though we found the switch for the security light, no matter if we turned it to the off position it would not go out. We even turned it off and left the area figuring that if it is controlled by a motion sensor, our movement may cause it to stay on. But when we came back to the area it was still on.

In the meantime, investigator Jamie had to leave, which left Jay, Chris and me in the basement and Tim monitoring his equipment upstairs on the stage.

Figuring that shutting the light off was a lost cause, we decided to continue our investigation in the basement by asking questions out loud, while using an SB-7 Spirit Box in hopes that our audio recorders would record responses made by disembodied voices.

For those not familiar with the SB-7 Spirit Box and how it works; The SB-7 is a mini AM-FM frequency sweeper, which is an altered radio that can fine-tuned to scan AM or FM radio frequencies, forward or backwards, at a rate measured in milliseconds.

The scan-lock mechanism on the radio is disabled. Therefore, the device continuously scans radio frequencies at a predetermined rate.

Sort of like twisting the knob on a radio backwards and forwards quickly, to produce random noise.

The original device was invented by amateur radio enthusiast Frank Sumption, who read an article about recording EVP that appeared in *Popular Electronics* magazine.

Sumption built a radio receiver he believed allowed real-time communication between the living and the dead and entities from other dimensions.

At first, I did not buy into the box working. I believed it was nothing more than radio static or skip coming through, along with a phenomenon known as audio pareidolia.

Audio pareidolia is the process of our mind, trying to make sense out of sounds and words and correlate what we hear to familiar words and sounds.

I soon changed my mind, as I noticed how specific information coming through the box were answering questions that I asked.

In addition, at the rate that I scan the frequencies, whole words and complete sentences should not be coming through when they are being vocalized by the same voice.

We conducted the session just outside of what was once Joe Neville's dressing room.

The session started off slow, with an occasional audible voice coming through, but nothing that was pertinent to the questions we were asking.

After twenty minutes or so, the voices that were coming through, were clearer and louder. At one point I asked, *"Is Joe Neville present?"*

Immediately we heard a voice say, *"Joe."*

We could not tell if the voice was saying that it was Joe or if it was a questioning voice asking, *"Joe who?"*

Thinking to myself, *"If this is Joe Neville and he is here in the building, maybe we can get a response to prove it is him."*

So, I said, *"If Joe Neville or any spirits are present, can you please turn off the light, because it is interfering with my video camera."*

As soon as I asked the question, the lights went out!

"That's not possible," I thought to myself.

So, I said, *"Thank you,"* and asked. *"Can you please turn the lights back on again for us?"*

Without delay, the lights immediately came back on.

We all looked at each other dumbfounded.

Even though he was standing three feet from the light switch, I asked Chris, *"Did you bump the switch?"*

To which he responded. *"No, I'm nowhere near it!"*

Hoping that the third time is a charm, I said. *"Just one more time, can you turn the lights off again please?"*

Instantly the lights went out.

This was unbelievable, we had tried every means possible to turn the lights off without success.

Now all I had to do was asked and they go on or off when requested.

Someone or something was manipulating the lights.

Even though I already said, *"Just one more time,"* when I asked for the presence to turn the lights out. I again said, *"Just one more time, can you please turn the lights back on again?"*

Well, immediately the lights turned on, but instantly turn off, then back on, then off again. This on and off sequence continued for close to a solid minute.

The lights turned on and off so fast that no one could possibly flip the switch that fast. They were basically flickering like a strobe light.

It was as if who or whatever was manipulating the lights were saying. *"Ok wise guys, I can turn the lights on and off all night, but what's the point?"*

If I had not been there to personally witness what happened with my own eyes, I would not have believed it.

The next morning, I called Scott Richardson and asked him if this had ever happened before.

He explained that not only had this never happened, the way that the lighting system is set up, it wasn't possible.

So, who was interacting with us? Was it Joe Neville, or was it some other spirit from days gone by who still roams the theatre?

I will probably never know in this lifetime, but the experience itself, ranks high on the list of the things I have witnessed as a paranormal investigator.

The next time you pass by the Legacy Theater at 101 East

Lawrence Ave in Springfield, stop by and take in one of the performances.

Feel free to strike up a conversation with one of the friendly patrons, but if you feel a cold chill and the hair begins to stand up on the back of your neck, you may want to take a second glance at the person sitting next to you... just in case!

The Monk Of Elkhart Cemetery

Over the years, we have heard legends of haunted houses and cemeteries and the stories of the strange activity that surround them.

If we decide to check them out for ourselves, we do so with a fair amount of skepticism and doubt as to the validity of the stories.

Elkhart Cemetery in Illinois is this type of place, so I decided to check it out for myself.

Located in Logan County, the graveyard is the final resting place of Richard J. Oglesby, the fourteenth Governor of Illinois.

Oglesby served as the Governor of Illinois from 1865 to 1869. When his first term ended, he practiced law until 1872, then ran for and was elected to a second term as Governor.

As it turned out, this was only a ploy and after his inauguration, turned the office of Governor over to his lieutenant governor in exchange for a seat in the U.S. Senate.

He served as a Senator until 1878. Then in 1884, was reelected governor for the third time. He was the first man in Illinois history to serve three times as governor. Oglesby retired and died at his estate in Elkhart and was buried in the Elkhart Cemetery.

My interest in the cemetery had nothing to do with politics

and everything to do with mystery and legend.

I had heard eerie stories of the deceased wife of Governor Oglesby seen visiting her husband's grave.

As the story goes, the ghost of Mrs. Oglesby sits outside of her husband's tomb. Shortly thereafter, a group of Native American Indian spirits approach her, and chase her over a bridge that leads to a nearby road.

If one legend was not enough for the graveyard, Elkhart Cemetery has a second one.

A fence separates the cemetery from a forested area that encompasses the property. On the other side of the fence is a path.

Legend has it that if you follow the path, you will hear voices and footsteps and see dark shadowy apparitions. You get the feeling that you are being watched and followed. At the top of a hill, the path branches, and splits so you must choose which way to go. If you choose the wrong path, legend is, you will not make it back.

In June of 2010, I was invited by a local paranormal group to investigate Elkhart Cemetery with them. I accepted their invitation, with the intent of finding out if the legends where true.

Especially the legend of the Native Americans chasing the spirit of Mrs. Oglesby out of the graveyard. If true, I hoped to get a glimpse of the event.

Accompanying the group was a former policeman familiar with the area and the cemetery.

We arrived at the graveyard thirty minutes before sunset. It was a clear evening and the temperature was in the upper eighties.

Upon arrival, we briefly walked around, then unloaded our gear.

I brought a limited amount of equipment, most of which was handheld, so it only took a few minutes for me to set up.

After I finished setting up my equipment, I stood nearby chatting with the team as they set up their equipment, in what would be our base camp for the night.

Nothing seemed out of the ordinary at first, but that would soon change, and I would witness the paranormal firsthand.

As the others set up their equipment, I was facing them and listening to the conversations taking place.

I couldn't put my finger on it, but something did not feel right. It felt as though I was being watched from behind.

Turning around, I visually scanned the tree line at the edge of the cemetery, hoping to get a glimpse of whatever was causing the feeling. But as I looked around, I did not see anything.

I know this can happen in a forested area, simply because you are secluded and surrounded by trees, but the feeling I had was different.

Usually once I get use to the surroundings, this feeling goes away, but it didn't, it became stronger.

When I turned back around and faced the group, the hair stood up on the back of my neck. It still felt like someone was watching me.

I continued listening to the conversations but could not shake the feeling of being watched.

The feeling was so intense that I turned around again and

looked toward the east but saw nothing out of the ordinary.

Turning toward the group I continued watching as the others finished setting up their equipment. The feeling I was being watched was stronger than ever.

So much so, goosebumps popped up on my arms and the back of my neck. It was so strong, I turned around again.

When I did, something caught my attention ten yards away at ground level.

I will never forget what I saw.

Squatting or kneeling was some type of being. I call it a *being* because I am not sure what it was.

I had the feeling it knew I could see it and wanted to remain hidden. It gave the appearance it was spying on us, or at least monitoring what we were doing there.

Although it was kneeling, I estimate it was three and a half to four feet tall and was dressed in a brown monk's robe with the hood pulled up over its head, looking directly at me.

The creepiest thing about it, was that even though I could tell it was looking directly at me, there was no face.

Where the face should have been was a grayish blur. All the facial features were gone, and where the face should have been was like a gray smear.

When I saw it, the first word out of my mouth was a swear word followed by, *"look!"*

When I yelled, the creature rose up and instead of walking or running away, it glided in the air for about ten feet, then vanished into thin air, disappearing into the twilight.

Whatever this thing was, did not want to be seen.

When I yelled out, the group looked toward me, but by the time they knew where I was looking, the creature had vanished.

What did I see? Was it a spirit, or was it something else? Why would it be dressed in a brown robe like a monk? I had never seen anything like this.

The way it glided away was like a witch flying on a broom. It did not appear to be a ghost but looked more like you would expect a ghoul or goblin to look like if such things exist.

It knew we were there; and was watching us. Which means, it was intelligent.

As I was explaining to the group what I had just witnessed, the ex-law enforcement officer that was with us, told me about an experience that he had in the cemetery on Halloween night years ago while he was on-duty as a police officer.

He told me about a grave in the cemetery where a man named Nicholas is buried. I will not give the last name of the deceased out of respect for any living relatives who may still live in the Elkhart area.

He explained that when Nicholas was alive, he was known to be a practicing warlock or male witch. As he told the story, he led us to Nicholas's grave. Which as it turned out, was only twenty yards from where I witnessed the monk like being.

As we gathered around the grave, he continued with the story.

"It was on Halloween night. I drove to the cemetery to make sure no kids were hanging out in the graveyard for Halloween. When I pulled in, my headlights lit up the area where Nicholas is buried. He explained.

"There was a group of people dancing around his grave. All were dressed in black and some appeared to be wearing hooded robes. I could not tell exactly what they were doing, but it appeared to be some sort of ritual. When they saw me, they took off running into the woods. I gave chase, but the group made a successful getaway."

He went on to explain that he was not sure what they were up to, but figured they knew Nicholas or had heard stories about him and were conducting some type of ritualistic dancing, using his graveside as a sanctuary of sorts.

The following Halloween, the officer checked the area around the grave, but never saw the group again.

As I mentioned, the area where I saw the apparition was approximately twenty yards from Nicholas's grave.

Could what I saw have been the spirit of Nicholas returning from the grave, or some type of being that may have been connected to him in some way?

But why had he been watching us? If it was Nicholas, was he curious of what we were doing or disturbed by our intrusion?

Or did it have nothing to do with Nicholas. Was it the ghost of a monk?

The former police officer told us that there had been rumors that due to the reputation of Nicholas being a warlock, his family had his remains removed from the cemetery and reburied in another location to prevent his body from being disturbed by those who practiced and worshipped dark magic.

We could see that something had been removed from his gravestone, as there was an outline of where something, had once been. The officer told us that at one time a picture of Nicholas was on the gravestone.

If stories of his body being relocated were true, it would make sense that his picture would have also been removed? But, why leave a headstone for an empty grave?

To me, it makes more sense that vandals, or someone who practiced witchcraft removed the picture from his gravestone.

In fairness to Nicholas and his family, I have not been able to substantiate the rumors that he dabbled in witchcraft.

But I will say that the former police officer, was adamant about what he witnessed on Halloween night years ago.

Those dancing around Nicholas' grave either believed the rumor to be true or knew something about him.

Later that night, one of the investigators took a series of photos near three small monuments. The stones were obelisk shaped and in one of the photos, she captured what looks like a large flame in front of the monument on the right.

Several unusual things were noticed about the flame in the photo.

We noticed that even though the flame is in front of the monument, the monument can be seen through the flame. When we examined the flame closer, it looked like there is a Cherub in the flame, or that the Cherub was causing the flame.

Cherubs are one of several types of angels that are mentioned in the Bible, Talmud, and the Koran. Most of us have seen replicas of them as statues, used in cemeteries and as lawn ornaments.

The image appears to be floating in the air in front of the stone and is three to four feet tall, which is similar in size to the being that I witnessed a few minutes earlier. So, did we capture a photo of what I saw that had transformed into the flame looking object?

The remainder of the night was uneventful, but how can you top witnessing a phantom in the first hour of an investigation?

We staked out Governor Oglesby's tomb to see if we could witness Mrs. Oglesby being chased out of the cemetery by Native Americans, but nothing happened.

We also found the path that the legend is based on, and even followed it. Unlike the legend, we managed to find our way back.

So, we were not able to prove if either of the legends were true or not. But we did have an encounter that the Hardy Boys and Scooby Doo would have been proud of.

We packed up at 1:45 A.M., as we all had to get up and go to work the next morning. But I had already decided to make a return trip to try and see the phantom monk again.

The following Saturday night, I returned to Elkhart to further investigate the graveyard.

Accompanying me was an investigator named Chris, who was part of the team from a few days before.

The weather conditions for Saturday evening were similar to earlier in the week.

Since the monk like being I witnessed and the strange photo that was taken, both occurred within thirty yards of Nicholas's gravesite, the logical thing to do was set up our equipment in this area.

If the phantom monk roamed this area, or if the phantom were Nicholas himself, maybe we could pique his curiosity and coax him out again to see what we were doing.

We arrived later than our previous investigation, so it was

already dark. We drove in far enough so we would be out of the site of passersby, then pulled off the main cemetery road and parked in an obscure location.

When I exited my vehicle, I had that same feeling of being watched as I did before, and I have been in enough dark and secluded places to get use to the feeling of seclusion. So, it was more than normal apprehension.

Maybe the being was watching our every move, so hopefully, I would see it again.

We unloaded our equipment, which consisted of digital audio and video recording equipment, with infrared capability for filming at night.

We set up chairs about fifteen yards from Nicholas's gravesite, which would be our base camp and observation point for the night.

Chris placed his video camera and tripod about fifteen feet from Nicholas's grave. The camera was equipped with an infrared light. Both the camera and light were powered by the cameras battery.

In addition to the built-in light source in his camera, Chris added an additional infrared light which also had its own independent battery source.

I set up an additional camera approximately twenty feet away, equipped with a built-in infrared light and additional infrared light as well.

I placed an audio recorder on top of Nicholas's gravestone. The device was showing low power, so I replaced the batteries with new ones.

Once everything was in place, we returned to our observation point to begin our stakeout.

We were having a general conversation as we monitored the equipment from our vantage point. The location we were sitting, was completely dark, making it easy to see and monitor the red infrared lights on our cameras, and the red indicator light on my audio recorder.

Hanging around dark places during paranormal investigations, reminds me of the old days as a private investigator conducting surveillances.

Many nights I waited for several hours and nothing would happen, while other times, things got interesting very quick. This night would be the latter.

Only five minutes into our steak out, something happened.

Suddenly, all the red infrared lights on our equipment including Chris's camera and external infrared light, the LED light on my audio recorder and the LED light on my video camera, all went dark.

Confused at what was going on, we looked at each other, then jumped up and hurried over to the grave.

We checked Chris's video camera first, and sure enough, it had shut down. His external IR light, as well as my digital audio recorder, video camera and infrared light all had turned off.

He checked his camera to see if his battery was still charged and it was completely dead. The external light, which operated on a separate battery system, was also dead.

If you think about the odds of this happening! Five separate devices, with fully charged batteries, all losing their power at the same time. It does not make sense.

No matter how you try to rationalize it, you cannot, it doesn't make sense logically.

There was nothing nearby to draw the energy from the batteries. A supernatural explanation makes more sense than the odds of a random coincidence of all the batteries going dead at the same time.

So, what happened?

Was there a spirit or some type of supernatural force in the cemetery that caused the equipment to shut down? Was it Nicholas, the alleged warlock, showing his displeasure of us recording by his grave?

Or was something else, possibly the phantom monk, making his presence known and showing us, he had the power to interfere with our equipment?

I have seen some unusual and unexplainable things in the years I have been a paranormal investigator. Things that make you shake your head, both in disbelief and in awe.

It seems the more I investigate the paranormal, the wider the door opens, and the more unexplained things happen. It is as though they know I am looking for them, so they come out to show me what they can do.

I will never know if the battery drain was intentional or merely the consequence of a spirit that may have come to close to our equipment.

Is there a possibility that something natural to the environment caused the power loss? I am open to logical explanations, but there was nothing around our equipment other than tombstones and trees. So, I am ruling out natural phenomena until someone can offer a better explanation to me.

There is a wonderful radio show that discusses and covers all things paranormal with which many of you are probably familiar. It is called Coast to Coast AM and is normally hosted by George Noory.

George has had guests on the show to discuss ghosts and hauntings. On several occasions, I have heard guests say that ghosts do not haunt cemeteries because it does not make sense for them to do so.

Most of the guest I have heard say this, are authors telling secondhand stories and have probably never spent a lot of time in cemeteries, or if they do, they do not go often enough to have encounters in these places.

I believe that when someone passes on, they still possess the free will that God provided them. I further believe that because of free will, some spirits choose to continue to have contact with the physical world.

But why would spirits choose to linger in cemeteries?

If you think about it, it makes complete sense for them to do so.

Because, if you are a spirit who wants to be noticed, what better place is there to get someone's attention than in the location of your final resting place, where family and friends come to pay their respects? Or where paranormal investigators like me come to record their voices?

Cemeteries are usually quiet peaceful places where a spirit would have a better chance of getting someone's attention than at the local mall.

What I witnessed in the Elkhart Cemetery was unexpected, and unexplained.

I often wonder what would have happened if I had not seen the phantom. Was it up to something or was it merely keeping watch on us?

If the stories about Nicholas are true and he was a practicing warlock. Is he earthbound, remaining in the physical world after death? If so, as a warlock, would he have an ability that allows him to do things that an average spirit could not?

Whatever I saw, was able to defy gravity. It lifted off the ground and glided away before vanishing.

Did I see a spirit, a witch, a ghostly monk, or some unknow creature or ghoul?

Although I am not sure, I do know that whatever it was, is one more reason that I keep investigating the mystery that Is the paranormal.

If I have learned anything at all as a paranormal investigator it is to *never say never* and never say *impossible*. Because as soon as you do, you will see something else that makes you shake your head and realize that it may be possible after all!

Ghost Children

On February 25th, 2012, I investigated a pre-Civil War era building at 406 West Gallatin Street in Vandalia, Illinois.

I was accompanied by a paranormal investigator who for this story, will be called Mr. Smith due to the nature of events that he experienced. Also, along on the investigation, was Sarah Hunter, the co-host of a local Central Illinois radio morning show, and her friend, Erica Lindsey.

The location is a paranormal hotspot with activity experienced throughout the entire structure. There are several businesses located on the first floor of the three-story building with the remaining floors relatively untouched and used primarily for storage.

In recent years, employees working at the building have reported seeing the apparition of a large man, as well as a little boy seen peeking around the corner of a work bench, in a flower shop located on the first floor

Staff have identified the man as a former tenant, named Bill Laswell.

Both the man and the little boy have been seen during the daytime, and the man has been seen at night as well.

This was my third investigation of the building and would turn out to be the most eventful to date.

The primary focus of our investigation this night was in a large room on the second floor.

There is a doorway in the room, that leads to a smaller room where Bill Laswell, had at one time lived.

We sat in chairs, staking out the large room for some time, as it offered a vantage point that allowed us to see into the small room as well as the hallway.

Shortly after 1:30 a.m., we decided to move our chairs from the large room and sit in the hallway so that we could conduct an EVP and a *spirit box* session.

I sat eight feet or so down the hall facing the doorway to the large room.

Erica was seated to my left and Sarah to my right. Investigator Smith stood leaning against a wall, facing away from the doorway of the large room.

We had been in the hallway for several minutes when I saw a flash of light coming from inside the large room.

Immediately, I spoke up explaining what I saw.

"I saw it too!" Sarah exclaimed.

Investigator Smith saw the light as a reflection on the staircase in front of him, that leads to the third floor.

Smith headed into the room and immediately announced that it was very cold. So, I followed him into the room and could feel the cold as well.

I asked Sarah and Erica to come into the room so they could feel the difference in the temperature.

When Sarah entered the room, she immediately yelled out and pointed toward Bill's old room exclaiming, *"I just saw a shadow move!"*

When I asked her what she saw, she told me she saw a large shadowy figure going into Bill's old room.

"Large shadowy figure? *That sounds like Bill Laswell,*" I thought to myself.

So, I headed for Bill's room hoping to get a glimpse of what Sarah had seen.

In the meantime, Erica stepped into the large room, so Smith, Sarah and Erica were all in the large room together.

When I walked into Bill's room, I did not see or feel anything unusual, so I turned around and headed back to the larger room to rejoin the rest of the team.

I took four or five steps into the room. Smith was standing in the doorway leading to the hallway and was taking photos with his digital camera.

The type of camera he was using, sends out a pre-flash that is a strobe light effect.

Due to the strobe light effect and camera flash, I could see where I was going without a flashlight.

Then Smith took another photo, and I saw them.

Standing directly in front of me, were two small children. They were no more than six or seven feet away.

When I saw them, I yelled out, *"I see two kids!"* I pointed to where they were standing and yelled for Smith to keep taking pictures, thinking that the pulsing light from his camera flash was allowing me to see the children.

Smith took another photo and the pulsating strobe light and camera flash lit up the room.

This time, I did not see anything, but immediately Sarah yelled out, *"I see them, they are holding hands and are running toward the door!"*

I froze and looked in the direction Sarah pointed but didn't see anything.

Then the camera flashes stopped.

I turned and looked toward Smith to see why he stopped taking photos. To my surprise his back was turned, and he was walking out of the room.

As soon as I saw this, I felt something was wrong.

Smith has a passion for the supernatural like I do and when something unexplained happens, like seeing an apparition, he would not turn his back and simply walk away.

I yelled out, *"Where are you going?"*

In a very slow, calculated voice and one that did not sound like his, he answered, *"I feel sad!"*

"You feel sad, sad about what?" I replied.

I called his name again. He did not respond and continued walking toward the wall of the stairwell leading to the third floor. Facing the stairwell, he put his arm on the wall and pressing his forehead against his arm, leaned against the wall, and stood there.

I moved toward Smith to see what was wrong.

Not realizing that after Smith snapped the last photo, he handed off his camera to Sarah by shoving it into her stomach.

When Smith did this, she realized that something was wrong, but continued taking photos.

Even though Sarah was new to paranormal investigating, and nervous about going to places reported to be haunted, she kept her wits about her.

Sarah's interest in the paranormal stems from wanting to know if it is real. Her curiosity seems to give her the courage necessary to overcome any fears she may have.

The photos she took would later help document what happened to Smith.

I walked through the doorway and approached Smith. Standing behind him, I placed my hand on his right shoulder and turned him so that he was facing me.

When he turned toward me, the reflection from my flashlight lit up his face.

Tears were streaming down his cheeks and all I could see were the whites of his eyes, as his pupils were rolled up into his head.

In a concerned voice, I said, *"Smith, what's wrong, buddy?"* I shook him trying to get him to snap out of whatever was happening to him.

Once again, in a voice that was slow, calculated and not his own, he said, *"I feel sad!"*

"What are you sad about?" I asked.

He started showing signs that he was going to pass out, so I grabbed him by his shoulders to steady him.

"Let's get you over to a chair so you can sit down."

He kept telling me that he did not want to sit down and repeatedly turned around looking behind him.

In a frightened tone of voice, he said, *"Who's behind me? There's someone behind me."*

I shined my flashlight behind him to show that no one was there.

Under no circumstance did Smith want to sit down. His eyes were still rolled up into his head and again looked like he was ready to pass out.

I knew I had to keep talking to him to keep him focused.

Finally, I was able to get him to one of the chairs in the hallway, but had to forcibly seat him, by placing my hands on his shoulders to hold him down.

At this point, his personality went from sad to angry. He seemed both frustrated and agitated.

Erica came over to where Smith was seated to see if there was anything that she could do.

I do not remember exactly what she said to him, but it was said out of concern for him.

But as soon as she spoke, Smith glared at her with an almost crazed look as if he wanted to hurt her.

So, in just a few short minutes, Smith's emotions went from sadness, to confusion, to anger.

After sitting in the chair for several minutes, he finally started to snap out of it and regain his senses.

The entire episode lasted little more than five minutes, but those five minutes were very intense and at times I was concerned for his safety, as well as the safety of the girls.

So, what happened to Smith? Was his body taken over by

spirits?

I do not think he was possessed, or at least not possessed in the normal sense.

What I believe happened to him that night, is this.

Prior to his odd behavior, he was standing in the doorway between the large room and hallway taking photos. When I walked out of Bill's old room and neared the center of the large room, Smith took a photo.

When the strobe light pre-flash and camera flash went off, I could see the two children, a boy, and a girl.

They were dressed like children would have been dressed in the days of *Tom Sawyer and Huckleberry Finn.*

So, they were probably children from the mid-1800s, which, is the time that the building was constructed.

The little boy was to my left and the girl was standing to the right of the boy.

Neither were moving and their faces were expressionless. They looked to be eight or nine-years-old.

I was looking directly at them, but they were turned at a slight angle and were staring directly at Smith.

What I saw reminded me of a scene from an old horror film. They had a blank stare on their faces. Their arms were hanging down at their sides. The oddest thing about them was that they were both colorless, including their clothing and faces. It was like looking at an old black and white photo.

The only exception was the little boy's chest, which had a pale-yellow splotch of color that really stood out and may have been the actual color of the boy's shirt. The sleeves of his shirt were either three-quarter or were slightly rolled up.

I am not sure why I had this feeling, but I knew they were brother and sister.

Their eyes were coal black. If you have ever seen an artist rendition of what are called black-eyed kids, this is what their eyes looked like.

I believe the two kids may have been following us around all night and thought that we could not see them. They were probably right, and we would not have seen them if it were not for the camera flash.

Somehow the strobe effect and camera flash allowed both Sarah and me to see them.

I believe that ghost or spirits, exist in a spectrum of light that we normally cannot see. But somehow, Smith's camera flash slowed down, filtered, or distorted the light spectrum allowing us to catch a glimpse of the two kids.

When I yelled out, *"I see two kids!"* They realized we could see them; they panicked and ran like you would expect scared children to do.

They headed for the closest exit, which they probably used hundreds and hundreds of times over the years.

Unfortunately for Smith, he was standing in the doorway and through their excitement; either one or both children ran through him.

One can only imagine what effect a spirit has on the human body when passing through it. Whatever happened to Smith, was very traumatic. Fortunately for him, he remembers very little of what happened to him.

Many paranormal investigators believe that ghost or spirits

are energy. I agree that spirits are energy, but if they are cognizant of their surroundings, then they are more complex than being simple energy, having thoughts and emotions as well.

In Smith's case, if one or both kids passed through his body, then what did he experience in a biological and psychological sense.

Sarah, Erica, and I, witnessed Smith's demeanor, personality and emotions change in an instant. During the five minutes or so that the event took place, it was as if Smith was a different person.

Did Smith take on the thoughts and emotions of one or both children? It certainly appeared that way.

Even though what happened seemed unintentional, it was a traumatic experience for all involved and was something that we could not control.

I am not trying to dissuade any of you who may be interested in investigating the paranormal by telling this story. But you need to understand that there are risks involved in investigating supernatural things. Things that we do not understand and cannot control.

Because of what happened to Smith, I no longer stand in entrances, exits or doorways of buildings I investigate.

Paranormal investigating is an interesting and exciting endeavor. But is one that needs to be taken seriously. If you approach it with respect and caution, you can minimize the risk. But if you let your guard down, bad things can happen.

Doppelganger

There are so many strange and unexplained mysteries reported around the globe. Things so bizarre, that even if we read them in a science fiction novel, would shake our heads in disbelief and laugh.

I had one such experience, in the spring of 2009, while I was in Arizona. To this day, I am still not sure what I saw, whether it was a time slip or the doppelganger of another person.

If you remember the old TV series, *The Twilight Zone*, you will remember host Rod Serling's opening monologue, which is so fitting to the story I am about to share with you.

The experience was so bizarre, I believe that I may have had my own experience with the real *Twilight Zone*.

Serling's opening monologue began.

"*You unlock this door with the key of imagination. Beyond it is another dimension, a dimension of sound, a dimension of sight, a dimension of mind. You're moving into a land of both shadow and substance, of things and ideas. You've just crossed over into the Twilight Zone.*"

My personal Twilight Zone story begins when my wife, Kathy, and I were in Chandler, Arizona, to watch our son, Cory, play baseball for his college team.

It was 5:30 A.M. the morning we were leaving to return home to Illinois.

I pulled into a large gas station, which had a number of gas pumps and kiosk.

I pulled up on the right-hand side of the island closest to the front door of the convenience store on the premises.

Getting out of my SUV I walked to the rear of the vehicle where the fuel compartment door was located.

I opened the door, removed the gas cap, and proceeded to place the nozzle from the pump into the fuel tank.

I turned around, facing the gas pump, to select the grade of fuel and begin dispensing the gas.

As I stood there, I noticed an older model Cadillac pull up on the opposite side of the kiosk, coming to a stop one pump ahead of the one I was using.

I watched as the driver's door swung open.

A very attractive young woman, who I estimated to be in her late twenties, step out of the car.

She was thin, very tall, and had long, shiny black hair. Between her hair color and dark golden complexion, I surmised that she was Native American.

She caught my attention for a couple of reasons.

First and foremost was the fact that she was incredibly beautiful.

The second reason was that she was wearing a very skimpy, two-piece denim bathing suit, which seemed completely inappropriate for the cool morning temperature.

When I exited the SUV, I noticed the temperature on the dashboard was fifty-nine degrees.

After the woman exited the car, she swung the driver's side door closed, took a step up onto the concrete island, hopped down off the island, then jogged toward the convenience store. She opened the door to the store and went inside.

I distinctly remember seeing, and hearing, the door of the vehicle slam shut after she exited it.

After watching her enter the store, I turned around to check the nozzle I had placed in the fuel tank, to make sure it was securely in the tank. Then depressed the trigger of the nozzle to begin dispensing gas.

This took three or four seconds at the most.

When I turned back around, I saw the driver's side door of the Cadillac, swing open, like it had done moments before.

I observed what looked like the same young lady, wearing the same two-piece bathing suit, step out of the car, swing the door closed, hop up on the kiosk island, hop down and jog over toward the convenience store, open the door and disappear inside.

"What the hell!" I said to myself, not believing what I had just witnessed.

Quickly, I finished pumping the gas, and proceeded inside the convenience store, to see if the woman, possibly a twin, for some unknown reason, exited the vehicle from the driver's side.

I looked around the store and for obvious reasons, she was easy to spot.

I observed her selecting food items from the pastry isle. But there was no twin, she was alone.

Next, I walked around the store, looking for a clone of the woman, but there was no one else that remotely resembled the girl.

After paying for my fuel, I returned to my vehicle.

I casually asked my wife if she had seen the woman get out of the Cadillac and go into the store.

She replied that she had and mentioned how the girl had to be cold in the outfit she was wearing.

I agreed, then asked Kathy how many times she saw the lady get out of the car and enter the store.

"*Once,*" Kathy replied, "*Why do you asked?*

I explained to Kathy what I saw, and she reiterated that she only saw the lady exit the car and go enter the store one time.

We waited and watched as the woman emerge from the store, alone.

She walked to the Cadillac and got in. As her vehicle pulled out of the station, I could see that she was the only person in the car.

So, what did I see, and was I supposed to see it?

Was the stranger somehow spiritually or supernaturally connected to me to the point I knew what she was going to do before she did it and saw her completing the act before it took place?

Did I witness a timeslip?

Did I witness her spirit, her doppelganger, or something else supernatural?

The incident did not seem to be a case of *Déjà vu*, in which a person has a strong feeling that an event taking place was experienced before.

Because the event I witnessed, took place twice, in a matter of moments!

Did I catch a glimpse of the woman's free will as she made a life changing decision about something she was contemplating and as a result an alternate life manifest into existence and I witnessed the event?

Did one of the dark-haired women vanish or did she rejoin her original self, undetected by me?

I am not sure what I saw that day or if I was supposed to see it at all. But whatever I saw, convinced me that the *Twilight Zone* is real.

Eye in the Sky
The Cumberland Sugarcreek Incident

This story is about an encounter that I experienced that to this day, do not know what happened.

Whether the event was ghostly or possibly otherworldly remains a mystery.

As you will see, the events leading up to the strange encounter are as bizarre as the event itself.

It began when I attended a paranormal meet up at Lincoln Land Community College in Springfield, Illinois.

The group is called the *Prairieland Paranormal Consortium*. Friend, and fellow paranormal colleague, Carl Jones, serves as the host for the monthly meeting.

At the time, a second group called the *Central Illinois UFO Group*, also held monthly meetings at the college, which is relevant to the story.

The Paranormal Consortium is an open forum for those interested in all things paranormal, while the Central Illinois UFO Group limits its discussions to UFO related activity.

During the meeting of Saturday, June 25th, 2011, Stan Courtney a local Bigfoot investigator, presented evidence of possible Bigfoot activity, between the communities of Glenarm and Chatham, Illinois.

Stan showed several photos of an eighteen-inch-long by seven-inch-wide footprint that was taken under an apple tree, on property, near Covered Bridge Road in Chatham.

The family told Stan that they had not harvested any apples, but the apples were always disappearing from the trees. They also reported hearing strange noises in their backyard.

Stan showed a second photo of a rabbit carcass laid out underneath the tree.

The gruesome thing about the carcass is the rabbit's head had been pulled off.

According to Stan, in many case reports of suspected Bigfoot activity, carcases of animals often appear on properties where fruits, vegetables or animals have been taken.

Researchers believe that this is a type of *gifting* from Sasquatch to repay for what they take.

In the fall of 2010 a co-worker at the Illinois State Board of Education where I worked at the time, and who lived near Chatham asked me to come out to the parking lot to look at a handprint on the rear window of her van.

She knew that I was a paranormal investigator and thought that I would be interested in what she had to show me.

The rear window of her van was very dirty because she lives in a rural area and must travel down dirt and gravel roads to get to and from her home.

She lives approximately eight miles from where the Chatham footprint was found.

The handprint was on the rear glass of her van. It was ten inches long and seven inches wide and looked like the handprint of a primate.

The features that stood out the most were the human-like

lifelines on the palm. It appeared to have long, hooked claws or fingernails. It was like nothing I had ever seen before. That is, until Stan showed a picture of the Chatham footprint.

The toes of the foot had the same long hooked claws or fingernails, exactly like the handprint on the back of the van.

During Stan's presentation, a member of the paranormal consortium spoke up and said that the night before, Friday, June 24th, a local paranormal group that he was with, was investigating the Cumberland Sugar Creek Cemetery located between Chatham and Glenarm, Illinois.

The group was seated in the cemetery, when suddenly something or someone began throwing rocks and dirt clods at them.

One group member was struck in the cheek and her face was cut.

Rock throwing is considered classic Sasquatch activity and is meant to scare off intruders from their territory.

Strange footprints were also discovered under a bridge on Covered Bridge Road. Stan analyzed the prints and believed them to be authentic.

As far as he knew, the people who found the tracks under the apple tree and the bridge, did not know each other.

A short distance from where the tracks were found, a man reported that his dog was startled by something in the forested area near covered Bridge Road. The man said for some reason his dog started hanging around the house, rather than running in the woods as he normally did.

During the meeting, Carl Jones mentioned how he had recently seen a round, spherical object hovering over a pond near the Glenarm exit, off Interstate 55.

The object was round and shiny. He could tell it was not an airplane or helicopter, because it did not have wings like an airplane, nor did it have a rotor blade like a helicopter.

It was there a few seconds then was gone.

Carl works at the College and was there the following Monday, June 27th.

During the middle of the morning, Carl and several co-workers were outside taking a break. While outside, he looked up and saw a cylindrical object flying in the sky.

He said the odd thing about the object was that it did not have wings. He pointed it out to three other people who were there.

This is where the story gets a little weirder. First, let me say that I have known Carl since 2006 and have always known him to be honest and straightforward, so when Carl tells me something, I believe him.

On Tuesday, June 28th, Carl was at work when he received a phone call from a fellow member of the Mutual UFO Network (*MUFON*).

The man told him that a wingless, cylindrical craft was seen in the sky in northern Illinois and was heading south toward Springfield.

He surmised that if Carl went outside, he might be able to see the object. So, he went out, but did not see anything unusual.

A short time later Carl went out to take another look, and sure enough, there it was.

He described the object as a wingless, cigar shaped craft, travelling in an easterly direction at a fast but consistent rate of speed. Carl watched the object until he lost sight of it.

Tuesday evening, June 28th, at the monthly Central Illinois UFO Group meeting, Carl discussed the sightings he recently witnessed, including the metallic sphere seen near the pond at the Glenarm exit off Interstate 55.

During the meeting, I sat next to a paranormal investigator named Chris and we discussed heading out to Cumberland Sugar Creek cemetery after the meeting, to see if we could figure out where the paranormal group was sitting, when they were pelted with rocks and dirt clods.

Neither Chris nor I had ever been out to Cumberland Sugar Creek cemetery, so after the meeting, Carl provided general directions to the cemetery for us.

I drove my car to the McDonalds restaurant off Toronto Road, near Interstate 55, a couple miles from the Lincoln Land campus.

Chris followed me and I left my car in the parking lot. We then headed to Glenarm in Chris's vehicle to see if we could find the cemetery.

I was not familiar with previous activity at Cumberland Sugar Creek, but later found out that for years, local authorities would stake out the cemetery looking for underage drinkers. Several times during the stake outs, authorities reported seeing strange lights in the cemetery.

Thinking that kids were causing the lights, the police would investigate, but could never find the source of the lights.

Chris and I found the cemetery without any difficulty. But since we came straight from the meeting, we did not have our investigating equipment with us. More importantly, we did not even have a flashlight with us, which is probably not a good idea when roaming about a dark graveyard.

We arrived a few minutes before 10:00 p.m. The moon was a waning crescent, which is roughly one sixth of a moon, so the cemetery was very dark. The sky was clear for the most part, which helped with visibility.

After exiting the car, we walked a short distance into the middle of the cemetery. In doing so, our eyes seemed to adjust to the darkness, making it easier to avoid obstacles, such as tombstones, that were in our paths.

Shortly after arriving, I glanced upward and noticed an object about the size of a bright star heading west to east. I pointed out the light.

Neither of us knew what it was for sure, but it did not have flashing lights like a normal aircraft would.

A short time later, we saw a similar light traveling from the southeast, and heading in a northeasterly direction. Seconds after that, we saw an identical light coming out of the south and heading north.

This time, I kept my eyes focused on the light. I noticed that both lights were headed in an upward trajectory, becoming fainter and fainter, then disappeared.

Then it hit me: the lights were not extinguishing; they were heading so high into the atmosphere that they were disappearing.

I do not know of any military or commercial aircraft that travels that high, so I was completely baffled as to what we were watching.

After the lights disappeared, our focus returned to the cemetery and our mission of determining where the paranormal group had been sitting when they were struck by rocks and dirt clods.

We had been in the graveyard for twenty-five minutes, so

our eyes had adjusted to the darkness and we could see our way around without stumbling over obstacles in our paths.

Several times Chris and I thought we saw movement in the cemetery but wrote it off as our eyes playing tricks on us.

The temperature was somewhere in the low eighties, but since we did not bring equipment with us, I could not determine the exact temperature.

Several times we walked into cold spots that caused icy chills and goosebumps.

Based on our conclusion from the layout and terrain of the cemetery, we headed to the fence line in the southeast corner of the graveyard because this seemed like the most logical area where the Springfield group had been sitting.

It was hard to tell in the dark, but the area was not as heavily camouflaged with overgrowth of brush as we thought it would be for something as large as a Sasquatch to hide behind, while throwing at someone.

On a second excursion to the cemetery the following week, we were able to determine in the daylight that beyond the fence line to the south, the terrain sloped down, making it possible for a large person or creature to hide and throw rocks without being seen, especially under the cover of darkness.

We were walking toward the south end of the cemetery and were approximately forty feet from the fence line to the east, and thirty to forty feet from the fence line to the south.

Chris was less than ten feet in front of me, facing south.

I was behind Chris, also facing the south, at an angle that was a step or two to his left.

We were discussing where someone or something could

have stood while remaining hidden from the Springfield group, when suddenly, a bright light lit us up like Christmas trees.

Even in the obscure darkness of the night, I could see Chris as plain as if it were high noon on a bright sunny day.

My first reaction was to turn my head and look behind me, thinking someone was behind us with a spotlight.

When I turned, my attention was drawn upward.

I tried to look up but was met by a blinding bright *yellowish-white* light.

The light was so bright, I had to extend my arm and hand toward it, to shield my eyes from it.

Later, I found out that Chris reacted in the same manner as I did.

I am not certain how high the light was above me, but it was at least eight feet above and no more than ten feet behind me.

It engulphed both of us and lit up an area around us of at least twenty yards or more. The blast of light lasted for about four to five seconds.

It disappeared almost as quickly as it appeared. There one moment and gone the next.

When I turned and looked up, all I could see was the bright light. It reminded me of the type of light used in the old spy thrillers where a blinding spotlight was shined into a spy's face to get him or her to talk.

There are no security lights in or near the cemetery. The only lights were a couple of small solar lights at the far end of the graveyard.

When I turned back toward Chris, he was pulling his arm down, as he had extended his arm like I had done to block the light.

Chris tried to speak, but the only word that made it out of his mouth was, *"What?"* He began to cough and choke and leaned on a tombstone between the two of us and was unable to speak for several minutes.

After his voice returned, he explained that he had a burning sensation in his throat and compared it to a bad case of acid reflux.

When we compared descriptions of what we saw and how we reacted. Our stories were identical.

I asked Chris to check his watch to see what time it was. In my research as a paranormal investigator, I read about similar events in which missing time occurred.

Chris's watch had a light on it, so he was able to see that the time was 10:25 p.m.

So, taking into consideration the three or four minutes that it took for Chris's voice to return, the light appeared sometime around 10:22 p.m.

Whatever the light was, made no sound.

This is what puzzles me about the incident.

We were lit up by a strange light in the middle of a dark cemetery. So, one would think seasoned paranormal investigators would conclude ghost or spirit related activity.

But from the very beginning, neither Chris nor I felt the light was ghostly at all. It gave more the feeling of being otherworldly, or inter-dimensional, but not spiritual.

If the light came from the ground, or even from the side, maybe it would have seemed more like ghostly activity, but it came from above us.

Our reaction to the light was also puzzling. We were initially excited about what happened. But our discussion about the light, and our excitement lasted only a few minutes.

It was like nothing unusual took place and we were disconnected from what happened.

We continued to look around the cemetery as if the incident never happened. But our excitement for the night was far from over.

Chris had his new cell phone with him, and after being in the cemetery in complete darkness for about forty minutes, he remembered that it was equipped with a flashlight.

He turned on the light and we continued our exploration of the cemetery.

We continued to the far southeast corner of the graveyard, where we found a mound of soft dirt piled next to the fence line.

Chris being a hunter, likes to look for deer tracks when on rural investigations, so he shined his light toward the ground looking for deer tracks.

Instead, he found three very large footprints.

Footprints in a cemetery are not necessarily unusual, but these were, because they were barefoot. We could see the toes and heel of the prints.

What drew our attention to the footprints was their size. I wear a size eleven and a half shoe and when I placed my foot next to the print, the footprints were three to four inches longer than my foot. This meant that the footprint was at least fourteen to fifteen inches long.

We found two similar footprints nearby, which were also barefoot. Chris took photos with the camera on his phone to document what we found.

The tracks appeared to have been there for several days, which meant whatever made them could have been in the area the night the Springfield group was pelted with the rocks.

In addition, we were only a mile or two from where Stan Courtney conducted his investigation and took the photo of the, eighteen-inch footprint.

After finding the footprints, we made another pass through the cemetery then decided to call it a night.

When we left the cemetery, it was like we had an epiphany and what happened to us finally sunk in. We were both equally dumbfounded and keyed up by what just happened.

As we discussed the event, we wondered if something otherworldly had taken a picture of us or scanned our bodies for some unknown reason.

Something shined a spotlight on us, but who and why?

Did it want us to know that it could see us? Was it trying to scare us? I did not get the feeling that something was trying to scare us off, or the light probably would have returned. Plus, we both remained calm after the incident and did not feel threatened.

I have learned from my experience as a paranormal investigator, when in places like cemeteries, haunted houses, or other dark secluded locations, my senses become keen and more focused.

When Chris and I were in the cemetery, my hearing was more profound, hearing the slightest sounds and noises. But I did not hear anything before, during or after the encounter with the light. It caught us completely by surprise.

After leaving the cemetery, Chris drove me back to the McDonald's parking lot where I left my vehicle. We discussed what happened for several minutes then headed our separate ways.

A few minutes later, I called Chris on my cell phone to continue our conversation. We were fifteen miles or so from each other on Interstate 55, heading in opposite directions.

Interstate 55 is in range of many communication towers, so we were in good reception areas during the phone call.

We had been talking without interruption, when suddenly; I heard what sounded like a distorted mechanical voice coming from Chris's end of the call.

This went on for about a minute. I kept asking Chris if he could hear me, but the strange mechanical voice continued to persist until I hung up and called him back.

My cell phone was an older model, so normally, if I were in an area that had a poor signal, my phone would simply drop the call.

But this call was different because I continued to get the distorted, mechanical voice.

When I called Chris back, he explained that he heard the same mechanical sounding voice as I did.

Later, after returning home, I told my wife Kathy what happened, then went to bed.

During the middle of the night, I woke up and was sweating like crazy.

It felt like I had a fever and even though our air conditioner was on, I was uncomfortably hot. I got out of bed and took my temperature, but it was normal.

The next day, I was sick to my stomach and for two days following the event, I was extremely tired and lethargic.

I spoke to Chris after the incident and he too was tired and lethargic following the incident.

Cumberland Sugar Creek Cemetery

Since the incident, Chris and I have discussed it many times. We both have concluded that the light source did not come from inside the cemetery and felt that whatever it was, was otherworldly? But did otherworldly mean extraterrestrial or interdimensional?

How did someone or something know that Chris and I were in the obscure darkness of the cemetery and what was their interest in us.

The object Carl witnessed hovering over the pond, was only two miles, *as the crow flies*, from the Cemetery.

He described it as being round and shiny, but I wonder if the reason Carl could see it in the daylight, was because it was reflecting sunlight. Would it have been visible at night and was the object that Carl saw, the source of the light?

Wednesday, July 6, 2011, Chris, and I returned to the Cemetery with our equipment and flashlights in hopes of having a second encounter with the strange light.

Unfortunately, there was very little activity.

I did have one strange experience at about 8:00 P.M.

It was still light out and I had been over near the fence line on the southeast corner of the cemetery.

Chris was about thirty yards away from me. I was walking toward him, and when I got within about twenty feet of Chris, I heard and felt what sounded like the wings of a large bird fluttering wildly right behind my head at about neck level.

I could even feel the breeze from the flapping wings. I turned around, arching my back, and pulling my face and shoulders away not knowing what was behind me.

It was as if I was trying to get away from something and expecting to see this large, angry bird behind me.

However, when I turned, the noise and the breeze stopped, and nothing was there.

I turned and looked at Chris, thinking he could see

whatever it was. But he had a puzzled look on his face, like he did not know what was wrong.

I started to ask him what had been behind me, but he responded before I could say anything. *"Don't ask me, there was nothing there!"*

When I told him what I heard and felt. He added, *"I was wondering what the heck was wrong!"*

He said that it appeared as though I was trying to get away from something but reassured me that there was nothing there.

The incident turned out to be the only activity that night, so we went home a bit disappointed that the light did not show up again.

The only answer I have to the question, *"What happened June 28, 2011?"* Is, *"Something did!"* But the bigger question of, who or what it was, remains a mystery.

My gut feeling tells me that it was not spirit or ghost related. But was some type of close encounter with something otherworldly. Either from outer space or some interdimensional realm.

But why Cumberland Sugar Creek Cemetery and was the light related to the Bigfoot type activity in the area?

John Keel who was an American journalist and influential UFO enthusiast, best known as author of *"The Mothman Prophecies,"* theorized that portals or interdimensional windows exist throughout the world that connect our reality to parallel dimensions.

He believed that the interdimensional portals, could explain how phenomena such as UFOs and cryptid creatures are witnessed but leave little tangible evidence to prove their existence.

So, are UFOs and Bigfoot interdimensional travelers, and is there a connection between the two?

I believe that there is and mentioned this in my previous book, *"Chasing Shadows."*

If you conduct your own internet search by using key words, Bigfoot, and UFOs, you will find a multitude of eyewitness accounts linking the two.

Following the Cumberland Sugar Creek Cemetery incident, Chris, and I filed two reports with Bigfoot investigator Stan Courtney. One regarding the incident at Cumberland Sugar Creek and one related to a similar encounter at another rural Illinois location, Ridge Cemetery at Williamsburg Hill.

During the encounter at Ridge Cemetery, Chris saw an unidentified orange light float from the sky and descend into the wooded area surrounding the graveyard.

Later that night, we found a large barefoot track in an obscure location at the edge of the wooded area.

So, both incidents involved finding large barefoot tracks the same night we had close encounters with strange lights.

As a result of the reports that Stan filed with the "BFRO," *Bigfoot Research Organization,* in November of the same year, we were invited to be a part of a what the "Finding Bigfoot" television show calls a town-hall meeting, where witnesses discuss their encounters with Bigfoot type activity. Then after the interviews, the show selects one or two of the stories to be included in the show.

Unfortunately, the show would not let us mention the strange lights that we saw when the incidents happened. It seemed that the unidentified lights, didn't fit in with the shows premise that Bigfoot is an undiscovered animal or some type of surviving Neanderthal

After our interviews were over, the associate producer came to me and said that one of the cast members, had read my first book, "Chasing Shadows", where I discussed our encounter with the blinding light in Cumberland Sugar Creek Cemetery.

In the book, I mentioned my theory, of how I believe these bright lights are associated with inter-dimensional activity and could be portals or access points for cryptids like Bigfoot to enter our realm of existence.

I can't reveal the name of the cast member because he or she confided information to Chris and me with the stipulation that we would not reveal who provided the information to us.

They wanted to let me know they read my book and based on what they had personally witnessed, totally agreed with me that strange lights are associated with Bigfoot activity and also believe Bigfoot is an inter-dimensional being.

The cast member was accompanied by an outdoorsman who I will call Mike, in order not to reveal their identity. Mike works on the show behind the scenes and participates in investigations although not on camera.

The pair went on to tell a bizarre story of something they encountered, that convinced them, Sasquatch, is an inter-dimensional creature.

The incident happened in the state of Washington while they were on an investigation for the show. They were setting up their tents, in a wooded area, when something caught their attention.

Approximately fifty yards away, they saw a bright light which they believed was a lantern, and a group of people walking toward them.

When the group got within thirty yards of them, they could see that it wasn't a group of people after all. What they saw was a group of four, Sasquatch looking beings, two large and two small juvenile creatures.

Mike said, *"Larry, I can honestly tell you, that for the first time, since I have been investigating Bigfoot, I finally saw proof that they exist!"*

Then they realized that the light they thought was a lantern, was floating in the air next to the group, and wasn't a lantern after all.

Instantly and without warning, the light that was thirty yards away, was suddenly right in front of Mike and the cast member and it was blinding.

Then it vanished, and when it did, so did the creatures, they were simply gone.

If their story is true, and I have no reason not to believe them, it proves to me that Bigfoot is some type of supernatural or inter-dimensional creature and is not native to our realm.

As a result of the private conversation, Mike and the cast member investigated Williamsburg Hill with us in November of 2011, but unfortunately, all was quiet.

So, is the strange object that Carl Jones saw floating over the pond, the large footprints discovered on private property, the handprint on the back of my co-workers van and the blinding light and footprints Chris and I found in Cumberland Sugar Creek Cemetery, all related?

Well, they took place in the same general vicinity between Chatham and Glenarm, so the possibility is high that they are.

Something knew we were in the cemetery that night. But the question remains; what was it and will it be back.

Stranger in My House

There is an old saying, *"Everyone wants a ghost, until they get one."* Well, this is such a story, but it is not about something that occurred at one of the many locations I have investigated over the years.

Because it is about what happened in my own home, due to delving into the supernatural as I do.

Early on in my pursuit of the paranormal, I met a gifted psychic. When we met, she did not know me, yet knew everything about me, including that I pursue the supernatural as a paranormal investigator.

The first time we met, she gave me a bit of advice pertaining to investigating supernatural things.

She warned me that, *"Larry, you must be one-hundred percent sure that you want to delve into the metaphysical."*

Because as she put it, *"If you continue to do so, you will open a door, and not just any door, but a door to the other side!"*

She told me that, *"Once the door opens, you will see and experience things that you will not understand and once the door opens, it will not be easy to close."*

She further advised that, *"At some point, you will not have to go looking for spirits, because they will find* you."

As it turned out, she was right on all accounts.

My wife and I have been married for thirty-six years and have lived at three different locations.

We have lived in our current home since 1994, and never had anything unusual happen anywhere we lived until I began investigating the paranormal.

In 2008, I investigated a murder house in Iowa and after returning home, we immediately had unusual and what I would describe as negative activity, take place for a period of three months. To this day, things happen that seem to be related to the murder house.

We heard footsteps and saw shadowy figures moving freely about our home.

One of the most unnerving things that I experienced was hearing a whispery voice say my name, *"Larry"*, directly into my ear, when no one else was a round.

I heard the voice on four occasions, twice at home and twice at my office in Springfield. The reality that it was not my imagination, finally set in, when others around me, who had no idea what was going on, also heard the voices calling my name.

I truly believe that it was during the 2008 Iowa investigation that the door to the other side the psychic described, opened. Because since then, both Kathy and I have experienced unusual activity in our house on a regular basis.

In 2012, I was contacted by an acquaintance who knew I was a paranormal investigator.

Her son, who I will call David, had been experiencing unusual activity in his home for about a year.

She told me that she would have contacted me sooner, but David was worried that if someone found out he had a paranormal investigator looking for ghost in his house, he would be made fun of.

She explained that during the night, he heard footsteps in the hallway when no one was there. Many nights, his bed shook, and he heard what sounded like someone nervously tapping their fingernails on the bedpost of his bed.

The sounds were unnerving, but he could live with the noises.

But in recent weeks, the activity increased.

One morning, David's fiancée who I will call Sarah, spent the night.

In the morning she was in the kitchen making breakfast when she saw the reflection of a man standing behind her in the glass door of a cabinet. When she turned around, no one was there.

David's mother proceeded to tell me the reason that her son changed his mind about having someone investigate the house.

"David has decorative candles located throughout the house that are displayed but never lit." She explained.

"Last night, something awakened David in the middle of the night out of a sound sleep. While awake, he decided to use the bathroom. When he walked down the hallway toward the bathroom, he noticed a light coming from another room. As he walked closer to the light, he could see one of the decorative candles burning brightly."

She explained that seeing the lit candle scared David.

"He had seen enough and asked me to call you."

A few days later I met with David, Sarah, and David's parents at the house to conduct a pre-investigation.

They told me that it was their belief that the spirit haunting the house was David's deceased grandfather who I will call Andrew, who built the house.

David ask me if I would be able to determine the name of the spirit haunting the house.

I explained to the family that I was not sensitive to spirits, and the only way that I might come up with a name, would be if my audio recording equipment recorded a name being called out.

I further explained that I knew a very gifted psychic, who had done walk throughs of locations I investigated and in fact had come up with names associated to a location.

A week later, I met David and Sarah for the second time which was merely to gain access to the house for the investigation.

The following week, the psychic met with the couple and did the walk through. I had a previous commitment, so I could not be there for the walk through.

The day after the walk through, I received a phone call from Sarah. She wanted to know if I had told the psychic the grandfathers name, which I had not.

She proceeded to tell me that during the walkthrough the psychic had indeed come up with the name, *Andrew*.

I only saw David and Sarah two more times after investigating the house.

Once, a few weeks after the investigation to present the evidence I recorded during the investigation. I did not see them again until January of 2015, when I had a chance encounter with them at a Walmart.

Every summer since 2010, I take the host of a local radio station to an alleged haunted location to investigate. Then during Halloween, we do a live on-air recap and play audio that was recorded during the actual investigation.

In May of 2015, I contacted the owner of the murder house in Iowa, that I mentioned earlier, to schedule an investigation for June 23.

The investigation would be part of the radio stations 2015 Halloween show.

I have investigated the murder house five different times. On three of the occasions, strange things have happened to me both before and after the investigations.

Sunday, May 16, the morning after I had scheduled the investigation, I woke up, had breakfast, and decided to check Facebook to see what was going on in the world.

I noticed I had 27 notifications which was an unusually high number of notifications for my Facebook account.

When I clicked on one, I was shocked. It was a condolence message to Sarah, regarding the passing of David.

Immediately, I sent a private message of condolence to Sarah.

Within minutes, I received a reply from her, with her phone number, asking me to call her.

My intuition told me, because of my interest in the paranormal, Sarah was going to ask me questions related to the afterlife, which were questions I would not be able to answer.

That afternoon, I called Sarah.

She asked me if I had heard what had happened to David, which I had not.

She told me that he had taken his own life. Then as I suspected, she asked questions related to death and the afterlife, that I could not answer for her.

Questions like, "Will God give him a second chance in the afterlife?" "Will I ever see him again?" "Was it my fault?"

Of course, I could not answer any of her questions, but I told her what I believed.

I suggested that when she was ready, and the time was right, to call the psychic who did the walkthrough to see if she could offer any insight to her questions.

David passed on May 15 and in the days following his death, strange things began happening around our house. We heard glass shattering, but when we would check for the source of the noise, we could not find anything broken.

I saw the shadow of a person walk down our hallway, at a time when no one else was home.

On another occasion, we heard what sounded like a stack of heavy books, being dropped on the floor. We looked for the source of the noise and found nothing.

The week of May 25 became even more interesting.

On Wednesday May 27, I was at work. A lady named Marj was seated just behind me at her desk. It was exactly eleven o'clock in the morning.

Out of nowhere I heard my name, *"Larry,"* whispered. The voice came from directly behind me. I turned around expecting to see someone playing a practical joke. But the only person around was Marj who had her back to me and was working.

I said to Marj, *"Did you just whisper my name?"* Marj has heard the stories about the unusual things that went on at our office in 2008 after I returned from the murder house.

She looked at me and said with a nervous smile, *"No, and you better not be bringing that crap around here again!"*

That evening at home, I heard someone walking through the house, but no one was home but me.

Later when Kathy came home, we were sitting in our living room and once again heard glass breaking. We went to check it out. At first, we did not see anything, then I noticed that a large decorative wine glass, that sits in front of our fireplace was cracked down the middle.

The glass had not been cracked prior to the sound of the breaking glass, so I assumed that is what we heard.

Thursday morning May 28th, at exactly eleven o'clock, I was at work, once again seated at my desk.

Just like the day before, out of nowhere, I heard my name whispered.

I turned, and the only one behind me was Marj. Knowing how she would react; I did not say anything to her.

I remember thinking, *"Here we go again."* Because I had merely scheduled an investigation at the murder house and strange things were already happening.

Later that evening, I was at the computer in my home office listening to audio from a recent cemetery investigation. I was wearing headphones as I always do.

Kathy who has brown hair and was wearing a long sleeve black sweatshirt and jeans, came into the office to tell me that she was heading outside to water the flowers.

I acknowledged her, and she left the room.

I heard her go out the front door and the sound of the door closing behind her.

There is a window to the left of my desk that faces our front yard. As I continued listening to the audio, I occasionally glanced out the window to watch Kathy water the flowers.

I had just turned back toward the computer when I heard someone walking and stomping their feet in our living room.

My first thought was, how did Kathy get back inside so fast, because I had just seen her in the front yard, and besides, why didn't I hear the front door close behind her when she came in?

The stomping sound became louder and louder and I could feel the floor vibrating under my chair with each stomp.

"What in the world is Kathy doing?" I thought to myself.

Then the stomping got even louder. I could hear and feel it coming down the hall toward my office.

I took off my headphones and turned to my right just in time to see the backside of a short person, with long brown hair and wearing a black long-sleeved sweatshirt, as they passed by the doorway of my office and headed into our bedroom.

When I saw the brown hair and black sweatshirt, I was relieved that it was only Kathy, so I put my headphones back on and turned toward the computer to begin listing to audio again.

My relief was short lived because as I turned, something caught my eye.

Through the window, I saw Kathy in the front yard watering the flowers.

Thinking that we had an intruder in the house, I threw off my headphones and rushed into the bedroom but there was no one there.

The only other room in that part of the house is a bathroom just off the bedroom. There is only one way in and one way out, so I had him cornered, or so I thought.

I rushed into the bathroom expecting to have a confrontation, but no one was there.

An ice-cold chill overcame me as I stood there perplexed by the empty bathroom.

Because I knew what I had seen was not of this world.

What I saw, did not appear ghostly, it was as solid as a living person, but people do not just disappear.

Still in shock and standing in the empty bathroom, I tried to rationalize what I saw. But how do you explain something as bizarre as a person vanishing into thin air.

My thoughts quickly turned to the murder house and the upcoming investigation I was scheduled to conduct in less than a month.

In 2008 my world had been turned upside down by whatever followed me home from the house. Was it back, and what sinister things did it have in store this time?

The haunting in 2008 went on for months and didn't stop until I finally contacted the psychic who performed a spiritual cleansing.

Due to the extreme nature of the negative energy that was present, the cleansing lasted three hours

After the cleansing, the psychic warned me to never go back to the house.

I explained to her that my purpose in investigating the paranormal is to experience it, to better understand it and for that reason I could not promise her that I would not go back.

Sort of like the adage, *"If you're going to drink the water, you have to get it from the well."* She was not pleased with my response but understood.

When I saw the phantom visitor, I was looking directly at it, so there was no mistaking what I saw. But even at that, I still tried to convince myself that somehow Kathy must have been in the house.

So, I went outside and ask her if she had come inside for anything, and she had not.

If it was not Kathy, then who or what was it?

I thought about contacting the psychic to see if she could tell me who or what I had seen and why they were in my home.

But I knew that she would be upset with me once she found out that I was returning to the murder house for another investigation. So, contacting her was out of the question.

Since contacting the local psychic was not an option, I decided to get in touch with a man who I met in 2012 while investigating an infamous haunted house in Atchison, Kansas.

The gentleman's name is Tony and the house that I refer to is the Sallie House. Tony and his wife Debra lived in the house with their newborn son from 1992 to 1994. From the very beginning, they experienced strange activity in the house, including things like hanging picture frames turning upside down on their own, objects mysteriously bursting into flames and physical attacks to Tony by an unseen force.

I have investigated the house on three occasions one of which I investigated alone.

Each time I go to the house, I meet with Tony and Deb at their current home for a few hours before investigating their former home.

It was during these visits that I found out that Tony is an incredibly gifted psychic in his own right.

I have personally witnessed his abilities on several occasions, and they fall nothing short of amazing.

Tony was not sensitive to the spirit world prior to living in the house, so something that took place during the haunting opened his connection to the other side.

Tony is familiar with the Iowa house as well. After their experience with the Sallie House haunting, Tony and Deb formed their own investigative team in search of answers to questions that they had about the paranormal.

One of the places they investigated was the Iowa house and they too experienced the negative side that the house had to offer.

Friday morning, I sent a text message to Tony in hopes that he could shed some light on what I had seen and what it wanted.

In the message, I explained the weird activity that had been going on at our house. How I was seeing and hearing unusual things. I told him about hearing my name called, the odd noises and told him what I saw Thursday evening.

I further explained to him that I had recently been on several investigations and was scheduled to do another in June. But I did not tell him where I had been or where I was going, in order not to influence what he came up with.

I asked him to focus and see if he could provide any insight as to who or what I saw stomping down the hallway of our house.

By noisily stomping as it had done, I knew whatever this was, wanted to get my attention. But why?

Because of the similarities to things that occurred in 2008, I was sure Tony would pick up on a presence from the murder house or from one of my recent investigations. But I could not have been more wrong.

I had an investigation Saturday night, so I did not see Tony's reply until Sunday morning. When I read his message, I was totally floored by what he had to say.

Immediately, I knew who he was talking about.

Below, is the actual message Tony sent me on May 31, 2015 at 11:22 A.M. I have changed the names and location mentioned in the message to protect the privacy of those involved.

"Hey brother, from what I pick up, you were visited by a person to whom you have ties to. I'm picking up on a man in

his 30's. I am picking up that he has been trying to get people's attention for a few weeks. He needs a female to know that he is ok and that he will be with her again. I got something about, he will always be with her in Springfield. Not sure if that means anything to you. I feel this man knew you and how you investigate and thought you could get a message to a woman he was in love with very much, and with her I feel like he is telling me Sara or Sarah! He also wants to acknowledge a younger female and boy, kids maybe. So far, I am getting Rachael for a child that he is acknowledging. I feel like he left very quickly, a bit confused, he kind of felt he should have talked more about things. I guess things that were bothering him. Anyway, with this man, I am getting the name David. He has a similar bond with you Larry, baseball, he loved the game. And I see him wearing a Cubs hat! I feel that he really liked the Cubs, lol! I am going to send you a quick sketch of this man that I picked up on. I drew it quick, so sorry, not the best, but maybe it will remind you of someone."*

As I read the message, I could feel the goosebumps rise as a cold chill ran down my spine, because I knew exactly who he was talking about.

I later found out that what Tony said in the message including the information about David's children and his daughter's name were accurate.

In addition to being spiritually gifted, Tony is a fantastic artist.

When he sent me the drawing of the spirit, he was receiving the message from, it left no doubt who the message came from, because it was the spitting image of David, right down to his smile.

At first, I hesitated contacting Sarah, because I did not know how she would react to such a story or whether she would believe me or not.

"*But what if it was me on the other side trying to get a message to Kathy,*" I thought to myself. Who or whatever this supernatural being was I saw stomping through my house, sure went to a lot of trouble just to get my attention.

Plus, what if this one event was the sole reason and purpose for me being drawn to the paranormal. So, I had to take a chance.

I contacted Sarah and told her that I had something important to tell her that was relevant to David but was not something that I could discuss with her over the phone.

She told me that she would be at David's house that afternoon and I could stop by.

I remember thinking to myself on the way over. *"How do you explain to someone that you have a message for them from the other side?"*

When Sarah answered the door, she was not her usual bubbly self that I remember from our previous meetings, which was understandable based on what she was going through.

We sat down at the kitchen table and I told the story.

I explained the unusual things I had been experiencing the last few weeks and told her about what I saw in our house a few days before. I told her about hearing my name whispered two straight days at work.

Then I told her about Tony and his gift and read his message to her.

She cried with each piece of the message that I read, but her tears were interrupted by gentle laughter and smiles as if to confirm, that she knew the words were from David.

She looked at me when I finished, then gave me a hug, a hug that I could tell came from her heart.

She explained that she had not been able to function since David's death and that in recent days, other than taking care of her children's needs, spent all her time grieving.

She could not find the will to move on with her life due to all the unanswered questions.

But, after hearing the messages, she had hope. Hope and a belief, that she would see David again.

Sarah told me that the messages Tony was somehow able to relay from David, were exactly what she needed to hear.

Even though she believed what I told her was true, she needed a bit of reassurance that the message, was indeed from David.

So, she asked if I would contact Tony and see if he could contact David again, and ask him for a specific message that only the two of them would understand.

So, I sent a follow-up message and asked Tony if he could get something specific from David that only he and Sarah would know.

Once again, the actual message from Tony.

"Larry, I tried to ask for something only his fiancé would know. First, I heard him almost chuckle and in a sort of sarcastic way, say that he was a closet Cardinals and Packers fan, kind of like he did not tell many people that. Then I was getting a date of July 12, 2013 and the words we built together. This time Larry, he brought with him, two ladies, one in her 60's and one in her 80's, almost as confirmations of life after death for Sarah. The lady in her 60's is named Sadie and helped David in transitioning. The

older lady is named Catherine. Larry, he had a final message that made no sense to me, which is, he will always live in her heart, and someday they will be together again, scouts honor."

After I received the message, I called Sarah and read it to her. With each part of the message, she once again became emotional with a bit of laughter mixed in. Then with a joyous laugh said, *"It's him!"*

She explained to me that everything Tony said made sense to her. David was a staunch Cubs and Bears fan and loved to tease her about her rooting for the Cardinals and Packers, which were her favorite teams.

The date of July 12 was the date that they took out a mortgage on the house together.

The lady in her 60's who helped David transition to the other side was her mother who had recently passed, and the older lady was her grandmother, her mother's mother. Tony was correct with both names.

The final message in which David told Sarah that *"They would be together again someday, "scouts honor*," was the proof that she needed, because David had been an Eagle Scout. There is no way that Tony could have known any of this she said.

I have had a lot of sleepless nights since the 2015 incident in which I saw what Kathy and I now refer to as the *"Thin Boy"*, thinking about what happened, and why I was allowed to see it. Sort of why me!

I wondered why in 2012 David started having unusual activity take place in his house, including the lit candle.

Of course, I cannot prove my theory as to why the activity started all of a sudden, but here is my conclusion.

What if David's deceased grandfather had insight and knew what David would do in the future? If he knew what he would do, he would also know that David would need to get a message to the love of his life, from beyond the grave.

A message to reassure her that it was ok to continue living her life and that what happened was not her fault and maybe at the same time, offer a bit of reconciliation for himself.

Did the grandfather know about me and what I do? If so, he must have known that investigating the supernatural is not a hobby for me and that my commitment to finding answers about paranormal things, would make me the perfect person to introduce to David?

But how does a deceased relative on the other side, make this introduction possible?

What better way for the grandfather to bring a paranormal investigator and his grandson together than to haunt his grandson's home, knowing full well his daughter-in-law, David's mother, knew such a person, me!

He would also have known that I am not sensitive to spirits but know people who are, and that I would contact them looking for answers.

Maybe how it all came together does not really matter, but it proves that life is a puzzle. A puzzle that sometimes does not make sense, until we find a few missing pieces that fit together. And once we do, it all makes sense.

My encounter with the Thin Boy was amazing to say the least. But the story does not end there, because, on Wednesday evening, September 14, 2016, he returned.

Much like my experience in 2015, he had an important message for someone from the other side and it would be my job to deliver the message.

The Thin Boy showing up in 2016, blindsided me more than his previous visit did. I had come to terms with being the messenger for a spirit go between but did not expect it to happen to me again.

When you finish reading the story, you will see how the two experiences were similar based on how *"Thin Boy,"* used Tony and I to get messages from beyond the grave, to a loved one.

Like 2015, a series of events had to take place to pave the way for the parties involved to meet, for the plan to work.

This lining up of the players, proved to me, that there is something out there beyond our imagination, watching over us, and when necessary, nudges and guides us in a direction so that we meet someone for a reason. If I am right, then fate is not by chance, but by design.

Once again, names and identifying information of some of those involved have been changed, to keep their identity private.

Someday, when the time is right, they may wish to reveal their identity, to tell their remarkable story. But for now, the story of what took place, outweighs the names of the parties involved.

On July 20, 2016, I was contacted by a media personality named Lucy, who was trying to contact a Larry Wilson involved with an organization that she was interested in doing a story about.

I explained to her that I was not the Larry Wilson she was looking for and thought this would be the last time I would hear from her.

But as it would turn out, we would soon meet again.

Saturday night, August 6 around 9:30 P.M, my wife Kathy

was in our bedroom tidying up.

I was nearby in my home office writing when she entered my office with a puzzled look on her face.

"Did Cory just walk into his bedroom?" She asked, referring to our son.

"No, he is in the living room asleep on the sofa." I replied.

I noticed she had a puzzled look on her face, so I ask. *"Is there something wrong?"*

"Well, I saw what I thought was Cory go into his bedroom. So, I went to his room to talk to him, but his light was off. I turned the light on and walked in, but he wasn't there!" She said with a perplexed look on her face.

Since I began investigating the paranormal, we have seen and heard strange things around our house. Some of the odd things, concern Kathy, so when weird things happen, I try to deescalate her concern by playing it off as something explainable, which is what I did in this instance. Explaining away what she saw, as a shadow caused by a passing car.

I am not sure if my theory convinced her, but she halfheartedly accepted it and returned to what she was doing.

The following morning, I was again in my home office writing. Cory was across the hall in our bathroom, taking a shower.

Kathy passed by the office door and said, *"I'm going to Kroger's to pick up a few things for lunch."*

I was in the middle of typing a paragraph, so I did not immediately respond to her. But someone did.

I heard a deep male voice say, *"Ok, that's fine."*

Kathy heard the voice too, because she stopped dead in her tracks and peaked around the corner of the doorway to my office and asked, *"What did you say."*

"I didn't say anything." I replied.

"Someone did!" She responded.

"It was probably, Cory." I countered.

"It sure didn't sound like Cory." She said, knocking on the bathroom door while calling Cory's name.

"Did you just say something Cory? She asked. *"No, I'm taking a shower."* He answered.

Standing in the doorway to my office with a concerned look on her face, Kathy said. *"If that wasn't Cory, then who answered me?"*

"I didn't hear anything!" I replied. Telling a white lie, trying not to alarm her, even though I heard the voice too.

Wednesday evening August 17, at 7:15 P.M., I was in our living room watching television, when I decided to get a glass of water from the kitchen.

Cory was in the living room, programing an electronic device and Kathy was outside.

After filling my water glass, I turned around and started to walk out of the kitchen. When I looked up, I saw someone or something walking casually out of my office, across the hall, heading toward the bathroom.

What I saw, looked like a very thin teenage boy, somewhere between four and a half to five feet tall, wearing a black sweatshirt and gray pants. He had brown shoulder length hair. Immediately, I knew it was the same entity I saw in 2015 that I call, Thin Boy.

Just before getting to the bathroom, he, or it, turned toward me and ducked down, raising both hands in front of his face hiding it from me.

When he turned, I could see that the area in front of his face was blurred, while the rest of his body was clear.

The entity was not an apparition, but was solid, like a living person.

As I moved toward it, it disappeared out of sight. From my perspective, I could not tell if it walked through the bathroom doorway or through the wall.

I was less than twenty feet from it when I saw it. I do not know if it was from the shock of seeing the entity, but my body was chilled to the bone.

By the time I got to the bathroom, it was gone. Or at least, I could not see it, nor did it feel like there was a presence in the room.

Immediately, I thought about Kathy thinking she saw our son walk into his bedroom a few nights before. What I saw, had the same physical appearance as Cory, only it was much shorter.

Instead of Cory, did she see the Thin Boy and was the voice we heard last Sunday, him as well?

I sent a message to Tony in Kansas and told him what I saw, and how it was like what I saw in 2015.

I asked him if he could try and contact who or whatever it was to find out what they want.

After sending the message, I didn't hear from Tony. I knew that when he contacts spirits, it takes a toll on his body, so maybe he didn't want to deal with it. Several times while at his house, I saw bloody scratches appear on his face and arms, while he was channeling a spirit. Seeing things like that, makes you a believer in the supernatural, real fast.

Weeks went by and I still did not hear from Tony, so, I let it go.

In October, I received a surprise message from Lucy. The message said, *"Hi Larry, me again. I heard that you are a paranormal investigator. I am a reporter, and would love to do a story with you, if interested.*

Lucy wanted to tag along with me on an investigation to do a story. So, I agreed and took her on an investigation in late October.

On an average year, I receive several request from the media to tag along on an investigation to do a story for Halloween.

As fate would have it, on October 23, I received a message from Sarah, David's fiancé who in 2015, the Thin Boy had arranged to use me to get a message from David to her.

Message from Sarah

"Hi Larry! I hope things are good with you. I need another favor from Tony. I lost the house. The only thing that keeps me going, are David's messages from Tony. Will you please asked Tony if he can get through to David again? I am moving and want David to come with me. I just need to know if he will come with me."

Because I had not heard from Tony after I sent him the message in September, I hesitated to reach out to him. Finally, a month later November 26, I sent a message to Tony.

The following is the message that I sent to him.

Message to Tony

Hey Tony! Hope all is well with you, Deb, and the Boys. I hope you had a wonderful Thanksgiving. I wanted to touch base with you about a couple of things.

First, I have not seen the thin looking boy, since I messaged you back in August. I saw him August 17 and my wife saw him before that. He has the same build as my son, "thin," except my son is 5'-11" and what I saw was no more than 5' - 0" tall. I am not sure why he showed up, but we have not seen him since.

Remember when I messaged you a year ago May, when all the weird things were happening? I was hearing my name whispered, hearing noises, then saw someone stomping through my house.

I messaged you to see if you could shed some light on what was going on. You meditated and were able to communicate with the spirit of a man whose house I previously investigated, who took his own life a few weeks before all the strange activity started.

The man's name was David and you drew a picture of him and sent it to me. You were able to communicate with him, and had several messages for his fiancée, Sarah.

Sarah contacted me recently. Her message is pasted below for you to read. She has gone through some tough times recently, including losing the house she and David lived in together."

Sarah wants to hear from him again, because she is worried that David is stuck in the house and since she no longer lives there, he will no longer be with her.

"Tony, I know contacting the other side, takes a toll on you, so it is up to you, if you feel you should try to contact David again."

"Take care brother. Tell Deb and the kids hello for me." Larry.

I received the following message from Tony later that day.

"Hey Larry, when this Thin Boy you see is around, do you notice a correlation of meeting people? Like you run into new people or people you have not seen in a while? I ask, because, I feel this Thin Boy, could be a type of spirit guide to you and leads spirits toward you who want to communicate. Each time you have brought him up, I sense another spirit with him."

Let me ask you. Have you been in contact in recent months, with a woman, an almost beauty pageant type woman? Beautiful, carries herself very well, confident. Someone in the media, a Lucille or Lucy?

I ask because the spirit I am picking up on who is with the Thin Boy, wants to acknowledge this woman. I get the spirit of a man in his seventies. Passed maybe five or six years ago. I get the name James real strong. He wants Lucy to know that he is very proud of her. What she has made of herself and sends love to a Jane or Janey. Say's he is with them at times and mentions something about petting his friend Missy.

"Anyway, I feel you've crossed paths with this Lucy person and through your spirit guide was able to bring through a relative to acknowledge her. As far as Sarah, I will get back to you later today."

After hearing from Tony, I phoned Lucy and explained to her about the recent activity that I had been experiencing. I told her about the Thin Boy and Tony and read the message from Tony to her from the deceased loved one named James.

As it turned out, everything made sense.

James was her Grandpa Jim. He was in his seventies and died five years ago, just as Tony said. Missy, the friend that James indicated he was petting, was a deceased pet.

Lucy was in tears as I read the message to her.

I also found out that Lucy was involved with beauty pageants, which Tony alluded to in his message.

After our phone conversation, I emailed Tony's message to her and told her if she wanted to ask him anything, to let me know, and I would contact him for her.

I received the following message from Lucy.

"Hi Larry!" I have not been able to stop thinking about this since I received the message. It is all very interesting for several reasons. A couple of months ago, I was watching Tyler Henry on TV and started researching mediums. I have never had any experiences with one but was looking for some kind of comfort for me and my family. I did not end up contacting a psychic but was open to the idea. You see, when I was a teenager, my cousin and aunt died in a car accident. It devastated my family and there is still a lot of pain, even after all these years. That is what I want you to ask Tony about. I am not sure what to ask exactly, but I would like to know if Grandpa Jim is with them. My Grandma was diagnosed with brain cancer this summer. That is when I first reached out to you by mistake wasn't it. It has been a challenging few months to say the least. I am sure that is why Grandpa has been trying to reach us. Thank you from the bottom of my heart for sharing this with me. I have been smiling through tears all day."

After Lucy's message, I sent a follow up message to Tony to see if he could contact Grandpa Jim again to answer Lucy's question about her cousin and aunt.

I received the following message from Tony.

"Hey Larry, I know it is late, I apologize. I thought I would try and focus on Lucy a bit before I fell asleep. I saw the image of a woman and she was embracing a younger girl. I get the name Christine. It felt like she wanted Lucy to know they are with each other and are better than okay. I heard the woman say, I am with my sweet Dana."

"Oh, asked Lucy if there is a reason, I see an image of a soybean. Feels like it is tied to her grandfather."

"Her grandfather says to tell Lucy, that he is fishing with Jesus."

"I also got the image of her grandfather and with him, was the silhouette of three men. He was acknowledging the three men. I sensed a bond between the grandfather and the male figures, like a brotherly bond but can't tell if they are really brothers or just close friends."

As it turned out, Tony was one-hundred percent accurate about everything.

Message from Lucy

"Larry my grandpa is with his three brothers who have passed. Wow, he is right on the money with everything. My cousins name is Dana, she was eight when she died. My aunts name is Christine. I love the part about fishing with Jesus. My dad and grandpa fished a lot together. The soybean makes sense as well. My grandpa worked at a soybean mill and the mill had a large picture of a soybean on the sign where he worked."

After my last conversation with Lucy, I didn't hear from her until I received the following message on Wednesday, January 11, 2017.

"Hi Larry, my grandma passed away Sunday overnight. She is the one who had brain cancer, which is why I believe my grandpa may have wanted to reach out to me. Sunday morning, I woke up after dreaming that my grandma and I were walking down a hallway. She hadn't walked since her surgery in August. I said, you know how much I love you, right? And she said, I love you too! Then I woke up and noticed that I had a text message that said. Grandma's gone."

"I just feel so blessed to have had the message from Tony and then have a dream like that. I know so many people who have prayed for a dream or a sign and never got it. I don't know why my loved ones chose to gift me with such special messages, but I am eternally grateful that they have! I am also grateful to you for helping connect me with them. The comfort you have helped bring to my life is immeasurable and I am so honored and blessed to know you and call you a friend. Thank you from the bottom of my heart!"

Although we occasionally hear unusual noises in our home, the thin boy has not been seen since August 2016. Will he be back? Only time will tell.

But there is one thing that I am certain of. For some reason, I was trusted to help a couple of spirits get messages to loved ones who needed to hear from them. As a result, it made a difference in their lives.

So, if helping two people cope and realize that the best of life is yet to come, is the only reason I became a paranormal investigator. Then the several thousand hours I have spent investigating the unknown, has been well worth it and I would do it all over again.

This my friends, is what makes what I do worthwhile and why I keep coming back for more.

Malvern Manor

One advantage of writing books about my adventures investigating the paranormal, has been meeting local radio and television personalities with an interest in the paranormal.

In 2011, I was invited to be a guest on a popular Springfield, Illinois morning radio show, *"The Morning Grind,"* hosted by Jason *"Bondsy"* Bond.

Bondsy, who has a skeptical interest in the paranormal, heard about my first book, *"Chasing Shadows,"* and invited me to be a guest on his Halloween show, to tell a few stories about my experiences as a paranormal investigator.

After hearing my stories, his healthy skepticism of the supernatural, led to my invitation, to tag along with me on an investigation, to see for himself, what the paranormal has to offer.

Well, as they say, *"The rest is history!"* Because, the first place I took him, things happened he could not explain.

Each year, since 2011, I take Bondsy and employees from the station, *"normally interns,"* on one or two investigations during the summer. Then on Halloween, we discuss the investigations and play unusual audio evidence that we record.

On August 9, 2019, I took Bondsy and interns Lauren and Kate, to spend the night in one of Iowa's most infamous haunted locations, Malvern Manor.

Taking Bondsy and his staff with me on investigations, has been rewarding for me.

Rewarding, by giving those who are not yet convinced the paranormal is real, the opportunity to experience the type of things I have witnessed over the last twenty years.

In a way, taking them with me is selfish. Selfish, because when I tell someone who has never experienced the paranormal about the unexplained things I have seen. They have no reason to believe me.

But once they see the strangeness that the paranormal has to offer, it offers validation to the stories I have told them.

Since, 2011, I have taken Bondsy and seven of his staff on nine investigations. Each of them, without exception, have witnessed things they cannot explain.

The 2019 trip would be no exception and would be a night that none of us would soon forget.

Malvern Manor was built in the late 1800s as a family run hotel. It prospered due to its close proximity to the railroad and the many traveling salesmen, who spent the night there.

Unfortunately, with the advent of the automobile, and the decrease in rail travel, the hotel closed.

In the mid-1900s, the manor was the personal residence to the Gibson family.

T.D. Gibson and his wife were not the biological parents of their children, as the children's parents were declared unfit to raise them. So, they were sent to live with the Gibson's who were their aunt and uncle.

The youngest child, Inez, was emotionally affected by the separation from her biological parents.

According to stories, one day, Inez told Otto her brother, she was going outside to jump rope.

Later that day, when Otto walked into her room, he found his sister with a jump rope around her neck, hanging from inside of the closet.

The cause of death was determined to be accidental, but rumors circulated that the true cause was emotional stress due to a drop in her grades.

Visitors to the home have reported hearing the voice of a child coming from her room and some believe it is the spirit of Inez.

After the days of the Gibson family, the building was converted into a combination convalescent home and minimum care facility, housing a wide array of patients, ranging from alcoholics to schizophrenics.

It is alleged, that due to the varied array of disorders, at times, the care the patients received, was lacking.

Some believe that the alleged lack of care, has led to the haunting activity taking place at the manor.

During the over six hour, four-hundred-mile trip, Bondsy ask me to tell a few stories of some of the unexplained things that I have witnessed over the years.

One story I told, was about the strange things that happened to me after I returned home from an investigation at another famous Iowa haunt, the Villisca Ax Murder house, which coincidentally is only thirty-eight miles from Malvern Manor.

During the story, I explained how I would hear a strange voice whisper my name in my right ear. The importance of this story will be apparent shortly.

We were scheduled to meet owner Kurt Fricke at 6:00 P.M. but arrived well ahead of schedule.

So, we decided to have supper at one of the local eating establishments.

While we were waiting for our food. Kate said to Lauren. *"Did you tell Larry what happened to you?"*

"No, I almost forgot!" Lauren replied.

Curious as to what happened, I asked. *"Why, did something happen Lauren?"*

"Yes," she replied. *"It was really weird."*

"Do you remember when you were telling the story about the murder house and mentioned how you would hear your name whispered in your ear?" She asked.

"Sure, I do." I replied. *"Why?"*

"Well after you told the story, I shut my eyes to take a nap. I know I wasn't asleep, when I heard a voice whisper, Larry in my ear. It was freaky!"

Since this type of thing has happened before. I immediately asked Lauren. *"Which ear did you hear the whisper?"*

To which she replied. *"My right ear."*

What Lauren heard, has happened to others, when I talk about the murder house. So, I believed her.

But hearing the whispery voice was only the beginning of strange things that Lauren would experience on this night.

When supper was over, it was time to meet the owner, so we headed to the manor.

Kurt greeted us with a handshake and a smile, then began a short tour of the manor.

The first section of the building he took us to, was an area known as the nurses wing. At the end of the hall leading to the nurses wing, is the old nurses station, with a counter and storage bins where the residents medical charts were kept.

At the convergence of the main hall and nurses' station is another hallway, known as the shadow man hallway. Named for Malvern Manor's shadowy phantom, often seen in the dark corridors of this area of the building.

Malvern Manor

Some who have experienced it, have felt a malevolent energy when it is present. According to Kurt, several female investigators visiting the manor, claim that a shadowy figure followed them out of the hallway and into the main part of the manor.

I am not sure why, but the nurse's wing and the shadow man hallway gave me an eerie feeling. It was the type of feeling that you are not alone, and someone or something was watching.

With this being said, it didn't take long before the weirdness started.

Bondsy, was recording video with his cellphone while Kurt was describing activity that has been experienced in the hallway.

Suddenly, he interrupted Kurt.

Bondsy interrupted him, because he saw an unusual white light, move in front of his cellphone, while recording.

He replayed the video, and we could see a bright light come out of nowhere and move in front of his cell phone.

The odd thing about the light appearing, is the doors to the adjacent rooms were closed. So, the hall was dark with no available source to cause the light.

I have reviewed thousands of hours of video from investigations and am familiar with light anomalies caused by dust, moisture, and pollen particles. With certainty I can say the light Bondsy recorded was not caused by any of those.

It was simply a white light that passed or flashed in front of his cell phone camera.

After the brief excitement we continued down the hallway to room seven.

To get to room seven, you turn right at the nurse's station, and it is the last room on the righthand side.

Kurt stopped us before entering the room, to tell a story about a very beautiful woman who was committed to the manor when it was a home for the mentally ill. Some believe that her spirit haunts the room.

The woman was committed at the request of her husband because she was causing physical harm to herself.

As the story goes, the woman was healthy and happy with a husband and children at home. She had long beautiful hair and seemed to have everything going for her.

Then one day, for unknown reasons, she felt that she was no longer attractive, and her husband did not love her anymore.

She became obsessed with her looks and stood in front of a mirror constantly brushing and pulling out her hair.

With no other choice, her husband was forced to have her committed.

During her confinement at the manor, staff would find the woman standing in front of the mirror in her room, combing and pulling out her hair.

Some paranormal enthusiasts who have visited the manor, have recorded a woman's voice in the room, while others have witnessed the door opening and closing on its own.

Later, you will read about two incidents that happened during our investigation, to intern Lauren. The incident took place at the opposite end of the hallway and may be connected to the spirit of the woman in room seven.

When our tour of the shadow man hallway was finished, we headed to the second floor to continue our walkthrough of the building.

I always approach investigations of locations where a fee is required with a healthy skepticism. But it was during our walkthrough of the second floor, I began to believe that Malvern Manor, was the real deal.

After climbing the stairs to the second floor, Kurt stopped our group and said that there was something he wanted to tell us, before the tour proceeded.

"Before we go any further," Kurt announced. *"I want to disclose that something will happen to at least one of you when we enter a particular room on the second floor, I guarantee it. To prove what I am saying, I will take one of you to the side beforehand, and tell them which room and what will happen, without telling the others."*

"This should be good." I thought to myself.

Kurt selected Bondsy as the one to confide the information to, then we proceeded with the tour of the second floor.

At the top of the stairs, the first room on the right side is a small inconspicuous room, known as Hanks room.

Kurt explained that a patient named Hank who had a penchant for violence toward women, including the nursing staff, had once lived.

I didn't get a feeling of anything out of the ordinary in the room, but without prompting, Bondsy asked intern Lauren.

"Why are you holding your stomach?" To which she replied, *"Because I have pains in my stomach!"*

"How long have you had them?" He questioned. *"They just started,"* Lauren replied.

"My stomach hurts too," added intern Kate. *"But I think it's because of the ranch dressing I had on my dinner salad that*

didn't agree with me."

After Kurt finished talking about Hank, we exited the room. When we entered the hallway, Bondsy asked the group to stop before proceeding with the tour.

He then asked the girls when they first noticed having stomach pains and both indicated the pains started as soon as they entered Hanks room.

"When Kurt pulled me aside, he told me one or more of us would come down with a stomachache in Hanks room." Bondsy explained.

"No way," exclaimed Lauren. *"No shit!"* added Kate.

We all looked at each other in disbelief because none of us, except for Bondsy, knew this beforehand.

It was at this moment, that I started to believe that the stories that Kurt was telling us, were true.

There were no chemical or moldy smells to cause the sudden stomachaches that the girls had, and if something environmental caused it, one would think it should have affected Bondsy and me, but it didn't.

The next room of importance that Kurt took us to, is one that intern Lauren would have several experiences, *"Graces room."*

Many ghost have been experienced at Malvern Manor over the years, but none who haunt the building are more famous than Grace.

Grace was a patient who lived at the manor much of her life. She not only suffered from schizophrenia but was afflicted with multiple personality disorder as well. It is alleged that she had as many as fifty-six different personalities.

Staff often heard a man's voice coming from her room chanting, *"The Devil is coming to get me,"* which was repeated over and over in a gruff voice.

But when they would enter her room, they discovered the voice was coming from Grace, as the man was just one of her many personalities.

Grace's sunglasses and wheelchair are still in her room, giving the feeling that she still lives there.

Later in the chapter, you will read about events that took place in her room during our investigation, that focused on Lauren.

Next, Kurt took us to the room, where young Inez Gibson was found hanging in her closet. While in the room, he told a story about a doll, that some believe is possessed.

The doll is known as, *"Number One."* It was given this name, because a paranormal team using the doll as a trigger mechanism to coax a response from the spirit of Inez, placed the doll on a dresser in her room.

The group claimed that the doll flew off the dresser into the center of Inez's room. When they reviewed audio from their recorder, they had recorded a voice they believe was the doll, saying, *"I am Number One!"*

The story doesn't end there. Kurt told us that more recently, people who have touched the doll have had bad things happen to them. He didn't allude to what happened, but said the doll is now kept in the manor office in a glass case.

He took us to the office to see the doll. Kurt refused to touch it himself but indicated that if any of the group wanted to hold it, we could take it out of the glass case and do so.

Bondsy, Lauren and Kate politely declined. But me, being

from the *"I have to see it to believe it,"* school of thought, said, *"Sure, I'll hold it."*

I felt confidant holding the doll, since Kurt held another doll, that he purchased from Costa Rica, that was advertised as being possessed. So, if Kurt held a possessed doll from Costa Rica, what harm could come from holding, *"Number One?"*

Well, I would soon find out. Because less than twenty minutes after holding the doll, I fell from the top attic step, to the bottom of the steps, causing a large bruise on my hip.

So, the question is, was the fall the result of my own clumsiness, or did *"Number One,"* play a role in my tumble down the steps?

The jury is still out for me on this question. But before the night was over, several odd things took place, that may or may not be connected to *"Number One."* which I will discuss a bit later in the chapter.

Introducing us to the doll, was the final part of the tour, so Kurt headed out, leaving the manor to us for the night.

The first thing we did, was set up audio and video equipment throughout the building, then headed to the nurse's wing and the shadow man hallway to begin the investigation.

Kurt's description of a shadowy figure following female investigators out of the hallway, and stories I read on the internet, portrayed the hallway as the center of the darkest activity. So, this part of the building was the logical choice to start our investigation.

Since the interns had never been on a paranormal investigation before, we decided to stay together for the first part of the investigation.

We headed down the first-floor hall to the nurse's station then turned left. On each side of the hall, where rooms where patients once lived.

At the end of the hallway was a locked exit door. In front of the door was an old bedpan and a cushioned seat or ottoman.

It is important to note, that the tile floor at this end of the hallway, was clear of debris. So, there was nothing to obstruct our path as we walked down the hall. The importance of this will be apparent soon.

We did not have an actual game plan for investigating the shadow man hallway, but after a few minutes of discussion between Bondsy and me, and a few more minutes to reassure the interns we would be nearby, we came up with a plan.

Based on Kurt's story of the shadow man following a group of female investigators down the hallway, it seemed logical that if the shadow man were to show himself to anyone, it would be to female interns.

So, after convincing the girls to be bait for a paranormal trap of sorts, our plan was to leave them at this end of the hall, while surveilling them, from the opposite end of the hallway.

We would only be one hundred feet away and the girls were in range of our flashlights. If anything happened, we could get to the girls in a matter of seconds.

After I brushed the ottoman with my hand to clear the dust off, the interns sat next to each other, with Kate to Laurens right. For reassurance, Kate interlocked her left arm with Laurens right arm while Lauren held a flashlight.

After confirming with the girls that they were ok with our plan, Bondsy and I headed to the opposite end of the hallway.

When we reached our destination, I shined my flashlight

toward the girls, and I could see them in the obscure lighting. It wasn't long, before the plan worked.

We could hear the girls whispering, when suddenly, Lauren called out in a nervous voice. *"Hey Larry!"* To which I replied. *"Is something wrong?"*

"Maybe," She responded. *"Something is playing with my hair!"*

"What?" Bondsy questioned.

"Something just pulled my hair!"

"No way!" Bondsy replied.

"Larry can you come down here?" Lauren asked, again in a nervous tone of voice.

"Sure," I said.

With that, Bondsy and I hurried down the hallway. When we got to the girls, Kate was acting a bit strange. She was laughing and crying at the same time.

When I asked Lauren, who has long hair, that was neatly pulled back in a ponytail, what happened. She said that she felt someone playing with her hair, then yank on it.

I shined my flashlight on Lauren, and saw a strand of hair, sticking straight out on the left side of her head, or the opposite side from where Kate was sitting.

It looked as though; someone had pulled her hair.

She said that she could feel someone touching her hair, just before it was yanked.

As you can imagine, the girls were a bit shook up. When

they calmed down, Bondsy volunteered to sit on the ottoman, while the girls and I went to the other end of the hallway, to see if anything would torment him like Lauren.

I noticed a small yellow plastic ball about the size of a baseball on the floor across from room seven.

So that no one kicked it by accident, I placed it in the corner behind us. You will understand the relevance of my action, momentarily.

As we stood in silence monitoring Bondsy, all was quiet. Then, Bondsy asked, *"Did you guys hear that?"*

"No, what did you hear?" I responded

"It sounded like a ball bouncing!" He explained in a perplexed tone of voice.

"No, I didn't hear anything," I said, moving toward Bondsy and shining my flashlight down the hall in his direction.

As I walked past the nurse's station, the beam from my light, lit up something in the middle of the hall, ten feet or so in front of Bondsy.

"Is that a ball in the middle of the floor," I questioned, pointing my light at what looked like a ping-pong ball in the middle of the hallway.

"No way!" Bondsy said.

I then heard, *"Oh my God,"* coming from one of the girls as they followed behind me, still a bit on edge from the hair pulling incident.

"It- is- a ball!" Exclaimed Kate.

As I walked closer, I could see that indeed it was a white

ping-pong ball.

One thing I am sure of, is there was nothing in the hallway when we walked from the ottoman to the opposite end of the hall a few minutes earlier, and now, sitting in the middle of the floor, was a ping-pong ball.

"That was not there before," Kate declared. *"No, it wasn't"*, Lauren added.

As we stood around the ball discussing how it got there, I asked Bondsy what the noise sounded like that he heard.

"It sounded like a ball bouncing, and now there is a ball in the middle of the floor that wasn't there before." He replied raising his voice in excitement.

I picked the ball up and noticed it was flat on the bottom like it had been stepped on.

So, there was no way that the ball rolled to the center of the hallway and no one from our group stepped on it, or we would have heard the sound.

"Ok," I said. *"It's my turn to set on the ottoman while you guys go to the other end of the hall."*

With that, Bondsy and the girls started walking to the opposite end of the hall.

I turned my flashlight on so that I could find the ottoman in the darkness, when I did, I noticed a shiny penny laying on the right side of the ottoman.

I remembered using my hand to brush the dust off the ottoman before the girls sat on it. So, I would have noticed the penny.

"Hey, do you guys remember seeing a shiny penny on the

ottoman?" I questioned.

"What!" Bondsy asked in disbelief. *"There was no penny there he continued."*

The trio walked back toward me as I stood by the ottoman, the beam from my flashlight, illuminating the penny.

"That was not there," Lauren said. *"No way!"* Added Bondsy.

"I bet it fell out of your pocket Bondsy," Kate proclaimed.

"It couldn't have," Bondsy retorted. *"I don't carry change. Remember back at the restaurant I told the waitress to keep the change. I don't like the feel of change in my pocket."*

After we all agreed that the shiny penny was not on the ottoman before, Bondsy and the girls once again headed to the opposite end of the hall, while I took a seat on the ottoman.

Several minutes passed, and I heard Bondsy and the girl's voices. It sounded like they were excited about something.

"Did you kick that ball Bondsy," questioned a giggling Kate. *"No, I didn't!* "He replied.

"Oh shit, here we go again!" Exclaimed Kate, in a nervous voice.

"What's going on?" I yelled out, asking the group.

"See that yellow ball over there on the other side of the hall?" Bondsy said, using the light from his flashlight as a spotlight.

"It just rolled out from behind us on its own and rolled down the hall."

When Bondsy shined his light, I could see the ball he was referring to. It was the same ball that just a few minutes earlier, I placed in the corner at the far end of the hall.

I joined the group as they were gathered around the ball. I asked them to show me where they were standing before it rolled.

They took a position three feet in front of the corner where I placed the ball. The floor is level, so it could not have rolled due to momentum.

If Bondsy or one of the girls inadvertently kicked the ball, they would have had to kick their leg backwards, a distance of three feet, to touch the ball.

For whatever reason, someone, or something we couldn't see, seemed to be manipulating objects and pulling hair in the shadow man hallway.

After taking a short break, we headed to the second floor to continue our investigation. Things were quiet, so we decided to head back to the first floor.

Kate and I were walking ten feet or so in front of Lauren and Bondsy, when I heard a loud bang that sounded like someone had kicked a closed door.

Immediately, I asked Kate if she heard the noise and she did. Bondsy and Lauren, did not hear it.

We checked the nearby rooms and didn't find anything out of the ordinary. But during audio review, I found two clear EVP's that were recorded shortly after the commotion.

The EVP sequence was recorded at 9:50 P.M. In the clip, you hear me asked Kate if she heard the noise to which she replies, *"Yeah!"*

As soon as Kate responds to my question, a male voice, whispers, *"No!"* Then, a second male voice says, *"Well Kate, you're not going to go prove it!"*

It's speculation on my part, but I believe the voice is telling Kate, that she is not brave enough to go looking for the source of the noise.

But regardless of what it may have meant, it knew her name.

Over the years, I have recorded my name being said, many times, even when investigating alone. Which means, the recorded voice is not the mistaken voice of another investigator calling out to me.

Saying our names, proves that the voices we record are intelligent and know who we are.

After heading back to the first floor, Kate needed to make a phone call to her college roommate, as class was starting in a few days after summer break.

Since there was no cellphone signal inside, Kate went outside to make the call and Bondsy accompanied her for safety sake.

While Bondsy and Kate were outside, Lauren and I headed back to the Shadow Man hallway, to continue investigating. Only this time, I took a video camera equipped with infrared capabilities.

We set up the camera near room seven at the opposite end of the hall and pointed it in the direction of the ottoman.

During casual conversation, I mentioned to Lauren that I was disappointed I didn't have a camera pointed toward her when her hair was pulled.

I was surprised when she volunteered to go by herself, and sit on the ottoman, while I monitored the camera.

It surprised me, because it is unusual for someone who has never been on a paranormal investigation, to volunteer to do something by themselves, let alone do so, after having an experience like Lauren did.

After walking Lauren down the hallway, she sat on the ottoman and I returned to the opposite end of the hall and monitored the camera.

While surveilling her through the camera's viewfinder. I noticed a shadow behind Lauren but couldn't tell if it was her shadow or the shadow of someone or something else.

I asked her to move her head back and forth, then zoomed in on her with the telephoto lens. I could see that it was her causing the shadow.

Shortly after zooming in, I noticed Lauren fidgeting then turn her flashlight on and look behind her.

Before I could say anything, she called out. *"Hey Larry, something is touching my hair."*

"You're kidding," I responded.

"No, it just did it again and now something is pulling on my ponytail"

Watching Lauren through the viewfinder, I zoomed in closer. No sooner than I did this, I saw Lauren's head violently snap back!

"Larry, can you come down here, something just yanked on my ponytail!" Lauren shouted, with an urgent but calm tone to her voice.

"Hang on, I'm on my way!" I replied as I hurried down the hall toward Lauren.

She was surprisingly calm for having her hair yanked by a phantom hand.

When I asked her to describe what happened, she said it was similar to when her sister yanked on her hair when they were children.

"I felt something touching my hair, just before it yanked on my ponytail." She explained.

Shortly after the incident, Bondsy and Kate rejoined us. After telling them what happened, we took a short break outside, then headed to Grace's room to continue the investigation.

According to owner Kurt Fricke, Grace's room is one of the most paranormally active in the entire building. So, we were not sure what to expect after all of the activity we had experienced in the Shadow Man hallway.

We were standing around, when Bondsy decided to sit in Grace's wheelchair to see if this would stimulate paranormal activity.

It wasn't long until Bondsy was overcome with an eerie feeling that he shouldn't be sitting in the chair.

He said that he couldn't put his finger on it, but he felt he needed to get out of the chair.

I asked him if he felt like he was disrespecting Grace by sitting in her chair, and he replied, *"Yes exactly!"*

As Bondsy got out of the chair, I thought to myself, *"Why not!"* Then, took a seat in the wheelchair to see if I got the same uncomfortable feeling, and I did.

It was the same feeling of being disrespectful to Grace, so I got out of the chair.

When I did, Lauren commented that she was getting uncomfortably hot, which didn't make sense because we had been in other areas of the house that were even warmer. Plus, we had turned on the air conditioning in this part of the manor a bit earlier, so Grace's room, was one of the cooler rooms we had been in.

No one else felt hot, so I shined my flashlight toward Lauren and noticed sweat was rolling off her forehead. Before I could say anything else, she became sick to her stomach.

"I don't know what's wrong, but my stomach feels terrible," she said. "It feels like I could throw up."

With that, we decided to head to the airconditioned kitchen so that she could cool off.

No sooner than we got to the kitchen, I noticed a long scratch mark on the left side of Lauren's neck. I pointed it out, and upon close examination, we could see that it was fresh and was approximately three inches long.

Then we noticed a similar scratch mark on the right side of her neck also fresh and about the same length.

"You probably accidentally scratched your neck with your fingernails," Bondsy pointed out.

"No, I didn't," Lauren replied. "Plus, I keep my nails cut short," she added, as she extended her hands showing us her neatly trimmed fingernails.

So far, all of the unusual activity seemed to focus on Lauren. But why?

Based on several conversations I had with her, combined

with hearing my name whispered during our drive over to Malvern Manor, I believe Lauren is sensitive to spirits and may be what is known as empathic.

An empath is someone who senses the emotions of those around them, to the point of feeling the emotions themselves.

If you think about it, the manor is a place where patients and staff probably experienced on a daily basis, the type of things that Lauren experienced. I'm sure there were times that patients felt sick to their stomachs and times when they scratched themselves or staff and other patients.

We were in the kitchen discussing the possibility of Lauren being an empath when I suddenly realized something.

At the opposite end of the shadow man hallway and only a short distance from where Lauren's hair was pulled, is room seven.

If you remember, room seven, is where a patient lived, who stood in front of a mirror, pulling out her hair. Could the phantom culprit that tugged on Lauren's hair be the ghost lady from room seven? Was she jealous of Lauren's hair and wanted to pull it out, or did she think it was beautiful and wanted to touch it?

I have been on a lot of paranormal investigations in twenty years, and the only time I have encountered someone getting their hair pulled, was at Malvern Manor.

What are the odds of this occurring in the hallway, where a woman's room is located, who pulled out her own hair? It makes sense, that if the ghost of the woman from room seven haunts her room, she would haunt the hallway as well and is still obsessed with hair even after death.

After discussing the scratches and the possibility of the woman being the culprit who pulled Lauren's hair, we took a break outside, hoping Lauren's upset stomach would subside.

We were outside for fifteen minutes, then returned to the kitchen area before continuing the investigation.

It was at this point that something unusual happened to me.

Bondsy approached me with a concern. He was worried that activity seemed to be escalating, especially with Lauren.

"First it was mild hair pulling, then a violent yank of her hair, then she felt sick, and now something scratched her." Bondsy said. *"Maybe we should wrap things up and call it a night."*

I am not sure why, but I shouted at Bondsy, which was out of character for me.

"You want to bring interns along, hoping that they get scared to make for good radio, then when things start happening, you want to go home. You need to start taking the paranormal serious and understand that it is not entertainment."

Bondsy tried to calm me down, but I walked away from him.

Did getting upset with Bondsy or falling down the attic stairs have anything to do with holding the alleged possessed doll?

A few weeks after the investigation, I had a conversation with Bondsy about this. When I asked him if he thought the doll had anything to do with my sudden quick-temperedness or falling down the stairs. He told me he had a conversation with the interns, and all had wondered the same thing.

Bondsy also indicated that he thought it was strange how

both intern Kate and intern Lauren, acted out of character as well during the investigation.

He reminded me of how Kate reacted when Lauren's hair was pulled. How she was laughing and crying at the same time. Which didn't make sense.

"On the other hand," he said. *"Lauren was totally calm during the night, even after having her hair violently pulled, getting sick to her stomach, and being scratched."* Bondsy explained that Lauren has more of an anxious or timid personality, so it didn't make sense to him that these things did not upset her.

We may never know if the doll had anything to do with the way we acted, but it sure makes me wonder.

Since we had not investigated the attic yet, I decided to head up there on my own, while the group continued investigating downstairs.

I was in the attic for an hour and nothing happened. When I returned downstairs things were calm as well, so we decided to call it a night.

The investigation was eventful to say the least. I don't think the interns expected anything would happen, so to say they got more than they bargained for on their first paranormal investigation, is an understatement.

EVP Evidence

The following day, I began reviewing audio and video from our cameras and recorders. Based on the personal experiences we had, I hoped to find evidence of paranormal activity and as it turned out, was not disappointed.

I should note that the only thing we physically heard during the investigation, was the loud noise recorded at 9:49 P.M. All other EVP's were recorded, but not heard by our team.

9:49 P.M.
The first EVP that was recorded, was the loud crashing sound that Kate and I heard while on the second-floor hallway.

The only device that recorded the noise, was a recorder we placed in a mostly empty room on the first floor that Kurt indicated was a good location for recording EVP's.

The best way to describe what we recorded, is that it sounded like something heavy toppled over or someone kicked a door. Upon checking the room where the noise was recorded and other nearby rooms, we could not determine the source of the commotion.

9:50 P.M.
The next EVP that I found, was discussed earlier in the chapter. After hearing the loud crashing sound on the second floor, I asked Kate if she heard it, to which she acknowledged she did.

It was during this time, that the audio recorder located in the first-floor room, recorded a whisper, followed by a voice that seemed to be interacting with the conversation Kate and I were having.

When I asked Kate if she heard the noise, and she responded yes. A whispery voice says, *"no,"* and for all intents and purposes, seems to be mocking my question.

Immediately after the whisper, a male voice says, *"Well Kate, you're not going to prove it!"*

Who or whatever is speaking, knows Kate's name and one could infer that the phantom voice is saying that, it knows Kate will not go looking for the source of the noise.

Somehow, the phantoms behind the voices I record, know who we are. Which leads me to believe that even though invisible to us, they are nearby and can hear our conversations.

10:46 P.M.
Unfortunately, only part of the fourth EVP of the night is distinct. The voice is male, and clearly says, *"Get,"* followed by what sounds like either *"her"* or *"them."*

The voice was recorded in the parlor. No matter if the voice is saying, *"get her or get them,"* the statement infers that it is instructing someone to get one of us or someone else.

10:55 P.M.
EVP number five, recorded at 10:55 P.M., and EVP number six recorded at 11:40 P.M., which you will read about shortly, both sound like the same voice, and sound like someone is disguising their voice.

Both EVP's were recorded in the room on the first floor that Kurt advised was a good location for recording unexplained phenomena.

The voice talks in a high-pitched voice reminiscent of the cartoon character, Mickey Mouse. The voice asks, *"Hey, did you hear that?"*

I am not sure what the ghostly voice is referring to, because we did not hear anything at the time it was recorded.

11:40 P.M.
The last unexplained voice of the night, was clear, and once again, sounds like the high pitched, Mickey Mouse type voice recorded at 10:55 P.M. It is a beckoning type voice that asks, *"Where are you?"*

Although it is only speculation, I wonder if the voices were that of Grace, mimicking one of her many personalities voices?

Malvern Manor was definitely worth the almost seven-hour drive from Central Illinois. Whatever is there, is not shy when it comes to interacting with the living and is a place that I hope to spend more time in the future.

I believe the reason Lauren was picked on, is due to her sensitivity to spirits. After all, if a ghost wants to interact with the living, who better to reach out to, than someone who can sense they are there.

But whether you are sensitive to spirits or not, doesn't seem to matter at Malvern Manor. Because even someone like me, who couldn't sense a spirit if they were sitting on my lap, can tell there is something ghostly lurking in the shadows and hallways of the manor. Something that for whatever reason, chooses to remain after death and interact with the living.

Malvern Manor was definitely a paranormal road trip, that I will not soon forget, and a place that someday, I hope to return to.

Epilogue

"The most beautiful thing we can experience is the mysterious. It is the source of all true art and all science. He to whom this emotion is a stranger, who can no longer pause to wonder and stand rapt in awe, is as good as dead. His eyes are closed."
Albert Einstein

The paranormal is a topic some folks cannot deal with and others choose to ignore.

I have come forward with the stories you have read, to set the record straight and to reassure those of you who have experienced the strangeness that the unexplained has to offer, that you are not crazy, nor are you alone.

The supernatural is a normal part of life, but understanding it is rare. Our lives are made up of two worlds. The one we see and the one just beneath it. Invisible, but there, nonetheless. A place where strange is normal and one that somehow has access to ours.

This moment we call life, has a plot and we are merely actors following a script that we may or may not be writing.

The paranormal can be both frightening and enlightening. On one hand, you don't want to open the closet door because it is dark in there. But you are dying to know what is inside.

Much like the inside of a dark closet, the places I investigate hold the answers to the secrets and mysteries of the paranormal.

What punched me in the back on that cold night in Anderson Cemetery? Was it a ghost or something more ghoulish?

When it comes to the paranormal, everyone has their stories, and the stories in this book are but a few of the strange things I have encountered over the years, that keep me coming back for more.

Is Elkhart Cemetery inhabited by a monk or possibly the ghost of a warlock?

Did some extraterrestrial space craft or alien being shine a spotlight on Chris and me at Cumberland Sugar Creek Cemetery?

Who or what is the Thin Boy, what is his mission and why did he choose me?

These are questions that I probably will never be able to answer. But are the type of wonderous mysteries that make paranormal investigating worthwhile.

I have concluded that I will never be able to unveil the answers to the many questions that the supernatural has to offer, and actually, I am not sure I am supposed to.

For me, it is about mystery, the quest and pondering what might be.

Truth is the only piece missing that makes the paranormal what it is. Once truth is revealed to us, the mystery is gone and the paranormal fades away.

I cherish the time I spend in my attempt to unravel the mysteries of the unexplained and continue to seek out and explore the things that go bump in the night.

In the end, we may come to find, that, "man and monster," are closer than we thought.

Even though not all of us are prepared for a shattering glimpse into the unknown, I hope this book piques your interest enough to set out on your own paranormal road trip of sorts, open the closet door and find out what is inside.

Happy Hauntings!

Larry Wilson

ABOUT THE AUTHOR

Larry Wilson spent a decade working as a private investigator, before turning his attention to the paranormal. He is the founder of Urban Paranormal Investigations in Central Illinois. In addition to investigating hundreds of locations throughout the Midwest, he is a "Best Selling Author" who has written several books on the topic, guest lecturer and has appeared on both television and radio programs.

He is founder of 11:11 Films, an independent film company that produces paranormal documentaries. Larry has also assisted in the filming of three paranormal documentaries for other independent film companies.

Wilson currently lives in Taylorville, Illinois with his wife Kathy.

For more information, please visit:
http://lwilsonurbanparanormalblogspot.com/

Like us on Facebook:
https://www.facebook.com/Urban-Paranormal-Investigations-327088597440791/

Photo by Kathy Wilson

BOOKS BY LARRY WILSON

Chasing Shadows
Echoes from the Grave
Dark Creepy Places
Where Evil Lurks
Dr. Ugs
Paranormal Road Trip

www.ingramcontent.com/pod-product-compliance
Lightning Source LLC
LaVergne TN
LVHW051558070426
835507LV00021B/2648